OXFORD GEOGRAPHICAL AND ENVIRONMENTAL STUDIES

General Editors

Gordon Clark, Andrew Goudie, and Ceri Peach

Editorial Advisory Board

Homelessness, AIDS, and Stigmatization

The NIMBY syndrome in the United States at the end of the Twentieth Century

Lois M. Takahashi

CLARENDON PRESS · OXFORD
1998

Oxford University Press, Great Clarendon Street, Oxford OX2 6DP
Oxford New York
Athens Auckland Bangkok Bogota Bombay Buenos Aires
Calcutta Cape Town Dar es Salaam Delhi Florence Hong Kong Istanbul
Karachi Kuala Lumpur Madras Madrid Melbourne Mexico City
Nairobi Paris Singapore Taipei Tokyo Toronto Warsaw
and associated companies in
Berlin Ibadan

Oxford is a registered trade mark of Oxford University Press

Published in the United States
by Oxford University Press Inc., New York

British Library Cataloguing in Publication Data
Data available

Library of Congress Cataloging in Publication Data
Takahashi, Lois.
 Homelessness, AIDS, and stigmatization: the NIMBY syndrome in the United States at the
end of the twentieth century/Lois M. Takahashi.
     p.   cm. — (Oxford geographical and environmental studies)
   Includes bibliographical references and index.
   1. Marginality, Social—United States.   2. Social isolation—United States.
   3. Homelessness—United States.   4. AIDS phobia—United States.   5. NIMBY
   syndrome—United States.   6. United States—Social Policy.   I. Title.   II. Series.
   HM136.T25 1998   98–23022
   361.6'1'0973—dc21
   ISBN 0–19–823362–0

1 3 5 7 9 10 8 6 4 2

Typeset by J&L Composition Ltd, Filey, North Yorkshire
Printed in Great Britain on acid-free paper by
Biddles Ltd, Guildford and King's Lynn

For David

EDITORS' PREFACE

GEOGRAPHY and environmental studies are two closely related and burgeoning fields of academic inquiry. Both have grown rapidly over the past two decades. At once catholic in its approach and yet strongly committed to a comprehensive understanding of the world, geography has focused upon the interaction between global and local phenomena. Environmental studies, on the other hand, have shared with the discipline of geography, an engagement with different disciplines addressing wide-ranging environmental issues in the scientific community and the policy community of great significance. Ranging from the analysis of climate change and physical processes to the cultural dislocations of postmodernism and human geography these two fields of inquiry have been in the forefront of attempts to comprehend transformations taking place in the world, manifesting themselves in a variety of separate but interrelated spatial processes.

The new 'Oxford Geographical and Environmental Studies' series aims to reflect this diversity and engagement. It aims to publish the best and original research studies in the two related fields and in doing so, to demonstrate the significance of geographical and environmental perspectives for understanding the contemporary world. As a consequence, its scope will be international and will range widely in terms of its topics, approaches, and methodologies. Its authors will be welcomed from all corners of the globe. We hope the series will assist in redefining the frontiers of knowledge and build bridges within the fields of geography and environmental studies. We hope also that it will cement links with topics and approaches that have originated outside the strict confines of these disciplines. Resulting studies will contribute to frontiers of research and knowledge as well as representing individually the fruits of particular and diverse specialist expertise in the traditions of scholarly publication.

Gordon Clark
Andrew Goudie
Ceri Peach

PREFACE

THE central argument of this book is that the rise in community opposition across race, class, and region should be considered in terms of the changing social construction of stigma, that is, the ways in which people define those who are acceptable and those who are not. This argument, and the overall logic of the book, are premissed on two assumptions. First, the twin societal crises of homelessness and HIV/AIDS, especially with ongoing contraction of the welfare state, are unlikely to disappear or even decline in the near future. Second, as more and more communities have discovered the tools and effectiveness of community opposition, the siting of particular controversial human services (i.e. those associated with homelessness and HIV/AIDS) has become increasingly contentious. The problematic of the NIMBY (Not In My Back Yard) syndrome can be traced in part to the increasing prevalence of homelessness and HIV/AIDS across the United States, and the ability of neighbourhoods and municipalities to act on their rejection through the widely publicized and used techniques defining the NIMBY syndrome.

In exploring scholarship explaining the NIMBY syndrome, I have been surprised at the general lack of engagement with current social theory debates concerning difference. Too often, organized community resistance to controversial facility siting has been viewed as the selfish, exclusionary practices of intolerant residents who are concerned solely with the potential impacts on local quality of life or property values. And often, perhaps because of this conceptualization of NIMBY as selfish exclusion, when addressing organized opposition, planners and policy-makers have frequently defined the NIMBY syndrome as the realm of negotiation and mediation, even as communities move to define such conflict as issues of localized autonomy and unfair state pre-emption of local land use control. Researchers interested in the complexity of community opposition, geographers in particular, have tried to move the scholarly debate of community resistance to facilities and land uses from these more instrumental perspectives to frameworks involving the state, production, and identity politics.

In this book, I draw primarily from the disciplines of geography and urban planning. The combination of these two disciplinary approaches provides both conceptual and pragmatic ways to examine the NIMBY syndrome. Urban planning tends to focus more on the pragmatic issue of finding ways to overcome local opposition across cities, while geography has placed more emphasis on the conceptual understanding of community response and organized resistance. Using these two perspectives, I want to demonstrate in this book the significance of the intersections among structural change, stigmatization, institutions, and communities to understand the future of the NIMBY syndrome. This treatment is not meant as a comprehensive study of all the interconnections. But in using various methods (quantitative and qualitative) to try to understand the dynamics of the NIMBY syndrome, my overall hope is to show that the geography of community opposition and stigmatization is centrally important to our understanding of the changing nature of locational conflict.

Three particular themes underlie the arguments made throughout the book: the importance of economic, welfare state, and demographic restructuring in community response to homelessness and HIV/AIDS; the significance of the social construction of stigma for ongoing and future community response; and the role of institutions like municipal governments and the courts in defining and adjudicating local facility siting disputes. The importance of each of these themes varies throughout the book through conceptualization to illustration of community and municipal response; however, each plays an integral role in the characterization of localized response. For example, Part III illustrates these themes through two case studies: Chapter 6 analyses the arguments used to manage homeless persons and facilities associated with homelessness in the Los Angeles metropolitan area; and Chapter 7 explores the perceptions of persons of colour in Orange County, California, towards HIV/AIDS and homelessness, specifically how these two crises are perceived in the Latino and Vietnamese communities, and the reactions of these communities to persons and places associated with HIV/AIDS and homelessness. While one theme may take precedence in specific chapters, reference is always made to the other themes. For example, in Chapter 9 intergovernmental efforts for reducing the stigma of HIV/AIDS are discussed with respect to HIV education and prevention services in the context of the changing demography and epidemiology of HIV/AIDS, and intergovernmental and organizational conflict.

This book focuses on the NIMBY syndrome as it relates to human service facilities, particularly those associated with homelessness and HIV/AIDS. In so doing, I am not proposing a theory of the NIMBY syndrome which explains all forms of community opposition, nor am I arguing that all forms of the NIMBY syndrome stem from the same sources. That is, community opposition directed at environmental hazards, at prisons, at shopping malls, and at factories, may have sources and mechanisms overlapping but also distinct from the NIMBY syndrome directed at homeless shelters and group homes for persons living with HIV/AIDS. Although I believe that contemporary community opposition against all types of facilities and land uses emanates from a combination of structural dimensions (such as economic restructuring and demographic

shifts), institutional response (such as 'welfare reform' and intergovernmental conflict), and differential community experience with homelessness
and HIV/AIDS, what I am suggesting throughout the conceptual and
illustrative chapters of this book is that, in addition to these, community
response specifically targeted at homelessness and HIV/AIDS is also based
fundamentally in social and spatial relations of stigma. In this way, I aim in
this book to clarify the particularities of the NIMBY syndrome as it relates
specifically to human service facilities, but I also have the broader intent of
understanding community opposition more generally.

Part I assesses the temporal and spatial patterns of community response
to controversial human services and their users, the reasons for the contemporary production of homelessness and HIV/AIDS, and the literature
which explains the causes and responses to homelessness and HIV/AIDS. I
argue that organized community rejection is not solely a concern about
property values or changing quality of life, or only the prejudicial responses
of intolerant individuals. The argument is made for a perspective rooted in
the social relations of difference, set within a more general conception of
the structure of stigmatization in the United States. Chapter 1 focuses on
the changing patterns of homelessness and HIV/AIDS in the United States.
Chapter 2 provides the bridge between Parts I and II, by providing a
critical review of the relevant literature on community opposition to
controversial human service facilities.

The three chapters in Part II develop the issues raised in Part I by
developing a perspective of community response based on the social construction of stigmatization. Chapter 3 provides an overview of theories of
stigmatization with respect to homelessness and HIV/AIDS. This chapter
allows me to explore the similarities and distinctions between homelessness
and HIV/AIDS, two societal crises which emerged coincidentally in the
United States during the 1980s and 1990s. Chapter 4 develops and expands
this perspective by exploring the intersections between stigmatized persons
and places. Chapter 5 considers the social construction of stigma even
further by incorporating race and gender, two dimensions which have vital
implications for understanding the changing nature of organized community resistance. Although I believe that the perspective of socio-spatial
stigmatization is necessary to understanding contemporary response, I
do not claim that this framework is essentially definitive. Rather, this
framework provides a useful tool to assess facility siting controversies
critically, to analyse decisions made by municipal governments, and to
explain the differential response to facilities across neighbourhoods.

In Part III, the theoretical perspectives developed on stigma and community response are used to explore two case studies, one focusing on
regional controversies concerning homelessness in Southern California
during the 1990s based on published media accounts, and the second based
on in-depth interviews exploring the perceptions of persons of colour in

Orange County, California, of HIV/AIDS and homelessness. The latter case is especially important to the book. Communities of colour have increasingly engaged in locational conflict, but mostly over environmental hazards, and primarily in working-class neighbourhoods/enclaves. In this case study, the opinions of professional and middle-class Latinos and Vietnamese individuals provide an opportunity to assess the interactions of race, class, and culture within the social construction of stigma.

After a consideration of community response to human services associated with homelessness and HIV/AIDS, Part IV is devoted to exploring municipal response by discussing two policies which are influencing local government policies across California and the United States as a whole. Because homelessness and HIV/AIDS tend to be seen by the public, service providers, and policy-makers as distinct, the two chapters in Part IV illustrate the exclusionary and inclusionary responses by municipal governments in two separate discussions. In Chapter 8, the institutionalization of rejection is illustrated by the recent ordinance passed by the City of Santa Ana to ban camping outdoors. This chapter considers the implications of such an ordinance for facility siting and community response. Also considered are the efforts for inclusion and acceptance illustrated by the conflict and co-operation in intergovernmental efforts to provide HIV/AIDS education and prevention services (Chapter 9). These two examples illustrate the rejecting and accepting potentials of public policy and form the basis for the final part of the book: justice, stigma, and the future of human service facility siting.

With an understanding of the socio-spatial construction of stigma and its implications for community and municipal response, Part V explores issues of fairness and the likely future of community and municipal response to controversial human service facilities. Chapter 10 analyses the complexities of fairness with respect to varied degrees of marginalization associated with homelessness and HIV/AIDS. This chapter emphasizes the role of federal and local legislation (particularly fair share policies), the issue of uneven distribution and equity in facility siting, and procedural justice. Chapter 11 concludes the book with a reconsideration of my arguments for a perspective focusing on the socio-spatial construction of stigma for clarifying the problematic of facility siting.

A number of chapters in this book were originally written as scholarly papers and first appeared in academic journals. These papers have been revised to reflect the thrust of the book. The revisions were done to consciously build links between the chapters and to reflect the conceptual framework. Overall, my goal was to provide a cohesive overview of the relationships between the production of the societal crises of homelessness and HIV/AIDS, the significance of the socio-spatial construction of stigma in defining community response, and the varying nature of community and municipal response to human services.

ACKNOWLEDGEMENTS

THIS book was written with the help of many people. First and foremost, I thank Gordon Clark for his guidance during the early formulation of the book idea through to conceptualization and writing. He has been instrumental in steering my thinking not only about this book, but also in theorizing social phenomena more generally. Bob Lake and Laura Pulido read and gave constructive comments on many of the central chapters in the book. They contributed greatly to the book with their critical insight about community opposition and mobilization, about the processes and outcomes of such practices, and about race and identity.

Many others have read portions of the book as chapters or papers, or contributed their ideas in lively and enthusiastic conversation about the issues. I thank Randy Crane, Amrita Daniere, Sharon Gaber, Robin Kearns, Patrick and Nancy Shea Kopka, Robin Law, Dowell Myers, Stacy Rowe, Elizabeth Rocha, David Sloane, and Rob Wilton. Their comments and conversations were very useful both in their criticisms, but also in their encouragement in bringing the concept to completion. I especially thank Jennifer Wolch and Michael Dear for teaching me not only about homelessness, human service delivery, and community opposition, but also about how to complete big projects. I also want to thank Glenda Laws who, before her untimely passing, not only provided moral support but also paved the way for much of my conceptualizing about the NIMBY syndrome.

The analysis of the community planning process for HIV/AIDS education and prevention presented in Chapter 9 is based on research which was originally co-authored with Gayla Smutny. The research project was a collaborative effort and I am very grateful to her for her insight and tireless effort. I am also thankful to Nick Compin and Dan Miyake for their research assistance, and to Michael White and Doug Wiebe for their artful map development. The staff at Oxford University Press were extremely helpful in facilitating the book through its revisions.

Writing and researching this book was made possible by a number of organizations. The Department of Urban and Regional Planning, School of Social Ecology at the University of California-Irvine, provided a much-needed sabbatical break in which to begin the project, and the University of California-Irvine provided a fellowship and release from teaching responsibilities to complete the book. In addition, I would like to thank the organizations which provided financial support for earlier versions of the research on national attitudes (Robert Wood Johnson Foundation and the John Randolph and Dora Haynes Foundation), and on 'minority' attitudes (National Science Foundation, Program in Geography and Regional Science, SBR-9308857, and the University of California-Irvine).

I would also like to thank the persons providing me with moral and emotional support throughout this process (in alphabetical order): Susie Kralick, Marta Slionys, Alan Takahashi, David and Machiko Takahashi, Hanae Takahashi, Joanne Takahashi, Connie White, Don and Renee White, and Lois White. Without their support, this project would have been much more difficult. I especially thank David G. White, to whom the book is dedicated, not only for dealing with the stress of book-writing with good humour, but also for steering me in the right direction when I yearned for inspiration.

Finally, I would like to thank the following publishers for permission to reproduce papers originally published as 'The Socio-Spatial Stigmatization of Homelessness and HIV/AIDS', *Social Science and Medicine*, 45: 6 (1997), 903–14, and 'Stigmatization, HIV/AIDS, and Communities of Color', *Health and Place*, 3: 3 (1997), 187–99, both with permission from Elsevier Science Ltd., The Boulevard, Langford Lane, Kidlington OX5 1GB, and as 'A Decade of Understanding Homelessness: From Characterization to Representation', *Progress in Human Geography*, 20: 3 (1996), 291–310, with permission from Arnold Publishers, 338 Euston Road, London NW1 3BH.

None of the mentioned individuals or organizations should be held responsible for any of the arguments or information expressed in the book. All of these are the sole responsibility of the author.

CONTENTS

LIST OF TABLES

LIST OF FIGURES

Part I

Community and Need

1

Understanding the Rise in Homelessness and HIV/AIDS

COMMUNITY opposition is an increasingly common phenomenon across the United States. In Sacramento, the State of California's capital city, a Loaves & Fishes dining hall feeding 1,000 homeless persons each day encountered vehement opposition by local businesses. The privately funded charity was sued by the City, which accused it of zoning violations (Warren 1997). Local business persons were upset about what they perceived were the 'hordes of homeless people who litter, trespass, deal drugs and otherwise pollute the ambiance downtown' (Warren 1997: A3). Steve Cohn, one of the City Council members supporting the lawsuit argued that it was 'really a product of frustration. . . . They put their religious mission above everything else and . . . it blinds them to the impacts they have on the surrounding area' (Warren 1997: A24). Loaves & Fishes Board members, however, argued that the charity spends $40,000 a year to clean the surrounding streets and attributes the lawsuit to a 'broader societal hostility towards the poor' (Warren 1997: A24).

This example hints at the complexity underlying community opposition. This downtown Sacramento dispute, couched in the language and legal structures of land-use control, reflects deeply held beliefs about the constitution of homelessness and the nature of the threat to residents and businesses. There are three primary dimensions to this constitution of homelessness, which may also be applied to HIV/AIDS:[1] homelessness and HIV/AIDS as populations of persons affected; as social and physical conditions; and as bases of stigmatization. This first chapter focuses primarily on the first two dimensions, that of homelessness and HIV/ AIDS defined as populations of persons affected, and as social and physical conditions. Later chapters (Chapters 3, 4, and 5) will elaborate on homelessness and HIV/AIDS as bases of stigmatization. Thus, to begin to understand the reasons for, and nature of, community opposition to human service facilities, this chapter first outlines the nature and causes of homelessness and HIV/AIDS to decipher the public's characterization of the populations affected and the social and physical dimensions of these two conditions.

One argument about the NIMBY syndrome is that its rise can be directly traced to the social and spatial expansion of homelessness and HIV/AIDS across the United States. Not only have the two crises not abated, but the population affected by homelessness and HIV/AIDS has expanded, diversifying socio-demographically and dispersing spatially. Homelessness and HIV/AIDS are no longer critical only for metropolitan and 'inner city'[2] residents, but have affected individuals and households across divisions of race, class, gender, and residential location.

One disturbing aspect of contemporary patterns of homelessness and HIV/AIDS is that the population characterized as at risk of becoming homeless or HIV-positive is widening. Public sector strategies to address homelessness and HIV/AIDS have been unable to stem the tide effectively. Long-term deprivation and marginalization for those homeless and/or living with HIV/AIDS have become the norm, and eliminating homelessness and/or HIV/AIDS an unrealistic goal.

And while homelessness and HIV/AIDS are becoming increasingly interconnected, scholarly explanation of their social and spatial expansion and the public and policy responses to them have remained relatively distinct. Policymakers, human service providers, and the public have tended to view homelessness and HIV/AIDS as two separate issues. This public separation of homelessness and HIV/AIDS has resulted in varying governmental responses although similar public rejection.[3]

This chapter highlights scholarly research into the nature and causes of homelessness and HIV/AIDS. There has been a growing and increasingly sophisticated literature examining the character, causes, and possible solutions for addressing both homelessness and HIV/AIDS in the US, particularly by geographers and urban planners. This chapter summarizes this scholarly work through two interlocking components. One is the description of the patterns of expanding and diversifying homelessness and HIV/AIDS in the US. Here, the focus is on the character and causes of homelessness and HIV/AIDS—who has been impacted and the reasons why these populations have become homeless or HIV-positive/living with AIDS. The second is an argument for the integration of public response in an exploration of homelessness and HIV/AIDS, in terms of understanding both the rise in homelessness and HIV/AIDS and the lack of political will to address these two issues effectively.

Homelessness in the US: A Decade of Research

Homelessness in the US became much more acute and increasingly visible during the 1980s and 1990s. The homeless population both increased in size and dispersed in space, so that homeless persons no longer resided solely in inner-city Skid Row areas, but also lived in many non-metropolitan, suburban, and rural locales (Blau 1992). But while the crisis of home-lessness has deepened, policy responses have become increasingly punitive. At the federal and state levels of government, for example, there continue to be calls for cutbacks in welfare spending, tightened eligibility require-ments, and renewed efforts at welfare state reorganization, and at the local level, there have been expanding efforts both to criminalize homelessness through anti-camping ordinances and to prevent homeless persons from entering and staying in specific jurisdictions.

Research on homelessness has become increasingly sophisticated and comprehensive over the past decade. Since the mid-1980s, scholars have worked to define homelessness, characterize the homeless population and the needs of homeless persons for services and housing, and determine the sources of the contemporary homelessness crisis.

Defining Homelessness

One widely accepted finding among homeless researchers has been that contemporary homelessness is distinct from that experienced at the begin-ning of the twentieth century, and, more recently, from homelessness during the 1960s. The new homeless have been characterized as younger, better educated, more often consisting of women and families, having significant proportions of veterans, and consisting of greater numbers of racial minorities than in the past (Baker 1994; Bassuk 1991; First *et al.* 1988; Hoch and Slayton 1989; Kozol 1988; Rossi 1989; Snow and Anderson 1993).

Although much of the past decade's research has been devoted to char-acterizing and counting the homeless population, the definition of home-lessness remains a topic of ongoing debate. Many definitions of homelessness centre on a lack of housing or stable dwelling, the absence of shelter on any given night, or life on the streets and in emergency shelters (Burt and Cohen 1989*a*; Morse *et al.* 1992; Ropers 1988). Broader defini-tions of homelessness move beyond the physical presence of housing, and include the notion that individuals and families live in inadequate housing circumstances not by choice, but because of a lack of social and material resources necessary to obtain and maintain housing (Stoner 1989). This broader definition has also been used to include those individuals living in flophouses, shelters, missions, and in overcrowded conditions (Wright 1989).[4]

It is clear from these definitions that the notion of homelessness spans a continuum of deprivation, from life on the streets to overcrowded housing (Hopper *et al.* 1985; Austerberry and Watson 1986). The range in the possible definitions of homelessness has constituted an ongoing dilemma for both scholars and policy-makers, both because of the subsequent difficulties in characterizing the population and estimating its size, and also in terms of assessing the importance of homelessness as a policy issue.

In the early 1980s, commonly reported national estimates of the homeless population in the US ranged from between 250,000 to 350,000 (according to a widely criticized 1984 HUD study) to 3 million (according to the Washington-based Community for Creative Nonviolence) (Hombs and Snyder 1982; HUD 1984). Scholarly efforts began to shift from counting the number of persons homeless to estimating the number of homeless episodes experienced by the population. By the 1990s, estimates of homeless episodes ranged between 840,000 for a given year (Wolch and Dear 1993) to 26 million over a period of five years (Link *et al.* 1994).

The US Census attempted to enumerate the number of homeless persons in its 1990 Census procedures. Estimates must always be viewed with caution, however. US Census estimates of the homeless population, for example, have been purported to have missed between 59 per cent and 70 per cent of the homeless population in Los Angeles alone (Cousineau and Ward 1992). The US Census count, though widely disputed, does provide an indication of the spatial distribution of homelessness throughout the US (Fig. 1.1). The distribution of homeless persons according to this estimate indicates, for example, that the widely held belief that homeless persons migrate to mild climates does not hold since the population is distributed across mild and severe climates, in northern and southern regions, and across states with large and small metropolitan areas.

Although the definition of homelessness continues to be debated, and the estimates of the number of homeless people varies widely, there is little doubt that homelessness has grown in significance. As one measure of this growing significance, public opinion polls have indicated that the public has increasingly come into contact with homeless individuals. A 1989 New York Times/CBS News national poll, for example, found that 51 per cent of respondents had had personal contact with a homeless person and that 65 per cent supported increased funding to help homeless individuals (Toro and McDonell 1992).

Characterizing and Explaining Homelessness

The support often found in public opinion polls for increased public spending to alleviate homelessness can be traced in part to the growing acceptance that homelessness is not caused solely by the personal deficits of individuals, but is the result of multiple factors, largely structural in nature (Lee *et al.* 1992). Structural explanations of the rise in homelessness focus on two interconnected trends: rising economic marginality and shrinking affordable shelter resources (Wolch and Dear 1993). A growing literature has indicated that the expansion of homelessness in the 1980s and 1990s has been fuelled by economic restructuring, ongoing welfare

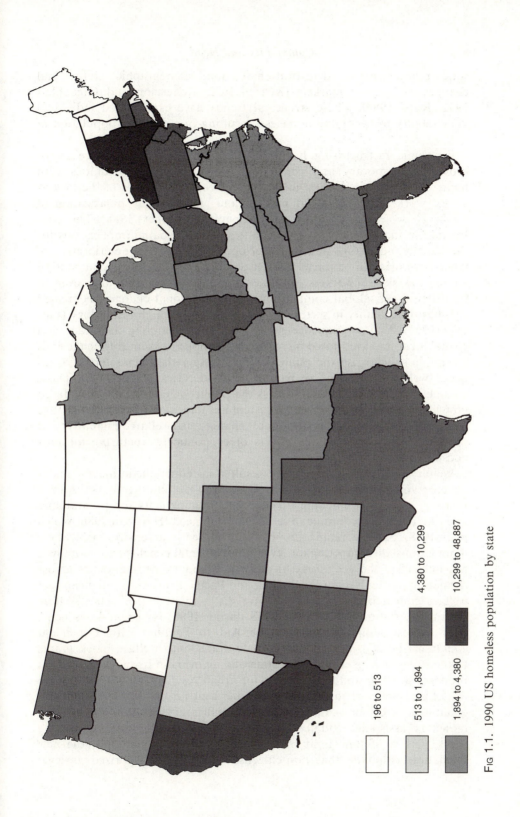

FIG 1.1. 1990 US homeless population by state

196 to 513

513 to 1,894

1,894 to 4,380

4,380 to 10,299

10,299 to 48,887

state reorganization, deinstitutionalization, demographic shifts, and changes in housing markets (Kearns 1992; Robertson and Greenblatt 1992; Rossi 1989). These structural trends have exacerbated individual vulnerability to becoming homeless, resulting in a growing population at risk.

Economic restructuring has resulted in an expanding service sector providing temporary and contractual employment opportunities with limited benefits, a reduction in the income-earning abilities of the lowest-skilled workers in the labour market, and an increasing polarization of wealth (Law and Wolch 1991; McChesney 1990; Ropers 1988). This contraction in income-earning opportunities has had spatial implications for the growth and change in urban, metropolitan areas. The deterioration of inner-city areas can be traced in large degree to the increasing competition among low-skilled workers for fewer and lesser-paid jobs, and, even in suburban areas, global competition and institutional change has created significant instability in sectors such as defence and high technology (Law *et al*. 1993). As a consequence, many individuals spanning class, race, and spatial location have turned increasingly to public assistance to make ends meet. Those who use the public welfare system risk characterization as being 'welfare dependent' or part of the 'underclass' (Gans 1995). Those who choose, and are able, to access the public welfare system have faced growing challenges as the welfare system has undergone significant change over the past decade. Institutional changes in welfare eligibility and provision has intensified the effects of economic restructuring for low- and moderate-income persons.

Welfare state reorganization has resulted in reductions in income maintenance and in-kind services, an elimination of funding for subsidized housing, and the tightening of eligibility requirements for programmes and funding which continue to serve people in need (Hartman 1986; Wolch *et al*. 1988). Such reorganization has contributed to an environment where more groups are competing for fewer governmental benefits and, moreover, has intensified the geography of poverty. Changing demographics across the nation particularly in the Sunbelt and Pacific Rim states will mean that welfare state reorganization will have distinctly spatial effects. 'Welfare reform' legislation passed in the US during 1996, for example, is being used to restructure the programme Aid to Families with Dependent Children (AFDC), known widely as 'welfare', into welfare block grants to the states (Lesher 1996). The State of California in its block grant plan proposed a cut in welfare payments for urban residents by 4.9 per cent (to $565 for a family of three) and by 9.8 per cent for persons living in rural counties (to $538 for a family of three). The federal legislation requires that monetary assistance end for able-bodied persons after two consecutive years, with a lifetime limit of five years for aid. In addition, the new regulations stipulate that non-citizens, even those categorized as legal residents, will not be eligible for food stamps or Supplemental Security Income (Richardson 1996).

But beyond ongoing efforts to restructure welfare programmes, policies from decades past have arguably been fundamental in producing the current growing need for housing and social services. One of the most significant has been deinstitutionalization, or the shift in mental health care provision from large institutions to community-based facilities. This policy, initiated during the 1940s and 1950s in the US, has often been blamed for the visibly increasing incidence of mentally disabled homeless persons (Dear and Wolch 1987; Mechanic and Rochefort 1990; Smith 1989; Torrey 1988).[5] At the very least, the growing number of mentally disabled persons with inadequate housing has led to increasing pressures for low-cost housing units and social services (Kearns and Smith 1993; Smith *et al.* 1993).

While many federal policy changes over the past decade have served to exacerbate the crisis of homelessness, some policy shifts have been responsive to research findings. For example, as researchers and policy-makers began to recognize that homelessness was not solely an emergency situation which could be resolved through short-term shelter, federal-level programmes began to include longer-term, permanent strategies for combating homelessness (Burt 1992; Kondratas 1991). Emphasizing this shift towards longer-term strategies, however, may paint an overly optimistic portrait of government response to homelessness. Large proportions of federal and state funding on homeless programmes remain focused on emergency measures, and policies such as the Stewart B. McKinney Act of 1987 suffer from constant underfunding (Blau 1992; Hopper and Baumohl 1994).

Also cited as a primary cause of the homelessness crisis is the massive demographic change which has occurred in the US, particularly over the past few decades. Significant changes in the structure of the family, for example, have been intricately linked to the increasing vulnerability of persons to becoming homeless. The feminization of poverty in particular has counted an increasing proportion of single women and female-headed households among the poverty ranks, and subsequently, at risk of becoming homeless (Kodras and Jones 1990; Ringheim 1990). Age has also played distinctive roles in the growth of homeless risk. The ageing of the baby boom generation has meant that more individuals are competing for jobs in an increasingly competitive labour market. This cohort has been subject to an expanding two-tier wage system which developed during the past decade because of the elimination of many middle-level management positions and the decline in the manufacturing sector (Harrison and Bluestone 1988; Phillips 1990). The risk of homelessness has also arisen anew for elderly persons in the US, due to the rapid growth in the size of the elderly population particularly over the past decade. Approximately 40 per cent of the single-person households in 1990 were over sixty-four years of age, indicating a significant population of elderly persons who are at risk of homelessness (Keigher 1991; Wolch and Dear 1993).

At-risk groups have been involved in an increasingly competitive housing market for affordable rental units, especially in metropolitan areas facing tremendous population growth pressures. Housing units which are affordable to low- and very-low income persons have been declining in number and quality, causing many to live in overcrowded conditions to remain housed and forcing others on to the streets (Baer 1986a; Burt 1992; Shinn and Gillespie 1994). An affordable housing crisis across the US has meant that fewer people are able to compete effectively in an increasingly competitive housing market. At a basic level, a shrinking housing supply for low- and moderate-income households, rising housing costs, changing demographics, and a downward trend in real wages have intensified the pressures for affordable shelter. Growing pressures on a shrinking affordable housing supply have created a housing market which resembles a game of musical chairs; the losers still standing when the music stops are often relegated to the expanding ranks of the homeless. The shrinking number and rising cost of housing units have been attributed to several developments: gentrification and urban renewal, demolition of single-room occupancy hotels, and the conversion of low-rent units into luxury apartments and condominiums (Baer 1986b; Stoner 1989; Hoch 1991).[6] Federal housing policy during the 1980s and early 1990s did little to counteract these market trends, rather moving to contract the state's role in the housing market by reducing budget appropriations and federally funded housing construction, and to incorporate market mechanisms such as vouchers (Bratt *et al.* 1986; Gilderbloom and Applebaum 1988).[7]

The emergency shelter system, often seen by the public as the housing option for homeless persons and families, remains problematic because of its dual and at times contradictory functions as (1) a housing option for those who are physically or mentally disabled, or without friends or family, and as (2) temporary housing support for those working individuals who are displaced from the labour market because of economic downturns (Hopper 1990). Rather than a co-ordinated system of housing, the existing shelter system is fragmented and at times does not serve those in greatest need (Laws 1992; Weinreb and Rossi 1995). Shelters have been generally associated with Skid Row areas of downtown metropolitan inner cities; however, there has also been a dispersion of specific shelter types into suburban areas. Specifically, shelters for women and children have encountered an increasing suburbanization, partly because of efforts at neighbourhood integration and partly because of pressures to move such land uses from expanding and redeveloping downtowns (Laws 1992).

The emergency shelter system cannot act as the sole option for transitioning homeless persons into mainstream housing. Many homeless persons choose not to use the existing shelter network because of the potential threats of robbery, physical and sexual assault, and the regimentation often dictated by shelter operators. In addition, researchers have argued that shelter life often creates significant obstacles to re-entry into housing. Termed 'shelterization,' homeless persons often encounter a cycle of recurring homeless episodes as they enter emergency shelters, are forced out on to the streets, and re-enter these shelters after a period of time (Hoch and Slayton 1989; Stark 1994).

While many geographers subscribe to structural explanations for the contemporary rise in homelessness, there are a significant number of social scientists who believe that individual vulnerabilities or deficits constitute the primary cause. Researchers argue, for example, that mental disability, substance abuse, criminal history, spousal abuse, family instability, veteran status, and other personal vulnerabilities lead to, or substantially intensify, homeless episodes (Caton *et al.* 1994; Drake *et al.* 1991; Gelberg *et al.* 1988; Hartz *et al.* 1994; Robertson 1991; Susser *et al.* 1991; Taylor *et al.* 1988; Weitzman *et al.* 1992; Wood *et al.* 1990). Some put this argument even more strongly, indicating that the homeless person her/himself, because of alcohol, drugs, and mental illness, constitutes the root cause of homelessness (Baum and Burnes 1993).

As these arguments indicate, there continues to be disagreement concerning the cause-and-effect relationship between personal vulnerabilities and homelessness. Research on the degree of social support available to homeless persons illustrates the ambiguities concerning how individual vulnerabilities may contribute to homeless episodes. Classic sociological studies on homelessness and Skid Row characterized homelessness as a state of disaffiliation, with homeless individuals (usually men) lacking community and family ties which would provide the resources necessary to remain housed and permanently employed (Anderson 1923; Bogue 1963; Bahr 1973). In recent comparisons of homeless and domiciled individuals, some studies have supported these earlier characterizations, that homeless individuals have smaller social support networks or that their access to social support is extremely limited (Cohen and Sokolovsky 1989; Hoch and Slayton 1989; La Gory *et al.* 1990; Solarz and Bogat 1990). However, the contrary has also been shown, indicating that homeless individuals have ongoing contact with family and friends (Goodman 1991; Shinn *et al.* 1991). Consequently, the lack of social support plays a vital but complex role in the paths into and out of homelessness.

Dealing with Homelessness[8]

Social support offers potential help by providing various types of support in coping with stresses and crises encountered in daily life.[9] Access to social support is a coping strategy used both by housed and homeless persons.

But for those who are homeless, access to social support networks may be critical in the differential ability to cope with extremely difficult emotional and material circumstances. The degree of coping with daily obstacles is measured for both homed and homeless persons by the degree of success that individuals have in meeting their daily needs (Rahimian *et al.* 1992). Homeless individuals who are better able to cope with their everyday circumstances may be able to obtain housing resources, to take advantage of employment opportunities, and to broaden social support networks (Rowe and Wolch 1990). Those with fewer coping skills may encounter deteriorating physical and mental health, and may be able to attain only shallow exits from homelessness (rather than becoming permanently housed). Coping strategies used by homeless persons include employment, involvement in the informal or illegal economies (such as day labour, selling blood, petty theft, prostitution, and panhandling),[10] support from family and friends, and the use of public and non-profit shelter and human service agencies (Takahashi and Wolch 1994). Such strategies are constrained by the social relationships among homeless persons, service providers, and social control agents, by the daily paths and routines of homeless persons, and by the organizational and locational characteristics of homeless service providers (Rowe and Wolch 1990).

The socio-spatial context of shelter and human services for homeless persons to a large degree defines the geography of homelessness. Research on homeless services can be divided into two major groups: those focusing on the service needs or demands by homeless individuals, and those focusing on the supply or provision of services. Studies focusing on the service needs of homeless persons have often emphasized the immediate and differentiated needs within the homeless population based on the length of time people are homeless, the cyclical nature of homeless episodes, and the socio-demographic characteristics of homeless individuals (Weitzman *et al.* 1992). In terms of health care needs, the rapid deterioration in mental health status, chemical and alcohol dependency, and the combination of such issues (resulting in the multiple diagnoses experienced by many homeless persons) have often constituted the primary focus for discussion (Belcher and DiBlasio 1990; Cohen and Burt 1990; Institute of Medicine 1988). However, also significant and receiving increasing attention are broader issues such as nutrition, sexually transmitted illnesses, traumatic injuries, developmental difficulties among homeless children and youth, and the desire to keep families intact (Bassuk and Rubin 1987; Cohen *et al.* 1992; DiBlasio and Belcher 1992; Weitzman *et al.* 1992; Wright and Weber 1987).[11]

With the many and varied needs exhibited by the homeless population, the system of traditional services has often been unable to respond in efficient and effective ways. The shelter and social service system in many cities contains what Wolch and Dear (1993) term gatekeeper services (points of entry into the human service system which usually focus on emergency shelter, food, and income maintenance services), coping services (health care, advocacy, meals, clothing, and other services which facilitate daily survival), and other human services (the remaining social services in any given system). Such services are provided by the public, non-profit, and private sectors, with the non-profit sector tending to dominate in terms of service delivery (Burt and Cohen 1989*b*). Although the need is great for housing, health, and other human services by the homeless population, several studies have indicated that the existing structure of services is underutilized (Herman *et al.* 1993). This underutilization may be due to the types of services being offered and the location of facilities and programmes.

Location in particular plays a significant role in the degree of access to services for homeless persons. The spatial distribution of facilities often coincides with the spatial boundaries defined by economic class. Consequently, more affluent communities tend to have facilities with universal access (such as libraries and recreation centres), while lower-income neighbourhoods tend to contain greater shares of the shelter and service system used by homeless individuals (Wolch and Dear 1993). While concentrations of facilities in specific locales may have agglomeration benefits for clients and facility operators, the location and composition of service-rich areas may also prevent many homeless persons from obtaining the services they need. That is, service-rich areas often do not contain, or are limited in supplying, all the services needed by individual homeless persons, leading either to necessary trips between service-rich areas or to attempts to cope without services. The location and composition of services are therefore critical, especially given the often limited mobility experienced by much of the homeless population (Wolch *et al.* 1993).[12]

The spatial distribution of shelter and services for homeless individuals has been constrained by federal, state, and local legislation, and has been influenced by community support or rejection of specific facility siting proposals (Laws 1992; see also Dowell and Farmer 1992). Municipal governments have long used the land-use planning apparatus (through zoning ordinances and restrictive covenants) to exclude facilities or limit their possible locations (Dear and Laws 1986; Mair 1986). In addition, municipal governments have increasingly turned to the criminalization of homeless activities to move homeless persons from their jurisdictions (Fischer 1988).[13] I will return to these issues in Chapter 2.

But homelessness is not the only social crisis gaining public attention during the past two decades. During the 1980s and 1990s, HIV/AIDS also gained prominence as a national crisis of public health, social norms, and public morality.

The Public Response to HIV/AIDS

The first case in the US of acquired immunodeficiency syndrome (AIDS) was reported in 1981. Fifteen years later there were well over half a million cumulative reported AIDS cases nationwide, with 40,000 additional people likely to contract HIV (the human immunodeficiency virus, believed to cause AIDS)[14] or AIDS in 1996 alone (Bennett and Sharpe 1996; Fig. 1.2). The devastating potential of this epidemic across multiple dimensions has been noted by researchers and policy-makers. HIV/AIDS has potentially far-reaching impacts on the nation's productivity. According to one analysis, HIV has become the leading cause of death for young adults (between the ages of 25 and 44) in many cities across the US (Selik *et al.* 1993). Public spending on HIV/AIDS had reached tens of billions of dollars by the late 1980s, with these costs including medical care, research, education, and blood screening (Earickson 1990).

By the early 1990s in the US, there had been almost 150,000 reported cases of AIDS, and estimates of between 800,000 and 1.2 million persons diagnosed with HIV (Gould 1993; Shannon *et al.* 1991). Spatial-analytic analyses of the diffusion and distribution of HIV/AIDS in the US have indicated that HIV/AIDS during the early 1980s was concentrated in major metropolitan areas in California, New York/New Jersey/Connecticut/ Rhode Island, and Florida, and has since spread outward from these core areas (Kearns 1996). Large cities in the US have become central in the developing characterization of HIV/AIDS. Reported AIDS cases in the metropolitan areas of New York City, Los Angeles, San Francisco, Miami, and Washington, DC, account for nearly one-third of all cases nationally (Table 1.1).

Defining HIV/AIDS and its Transmission

From tracking and analysing epidemiological information over time, researchers and policy-makers have concluded that HIV/AIDS can only be transmitted through sexual contact, exposure to infected blood or infected tissue (through needle sharing, infected blood products, or trans-plantation), or perinatally (either via the placenta or during birth). Following these conclusions, researchers, policy-makers, and service providers have attempted to dispel widely held myths concerning the potential for HIV/AIDS transmission via casual and non-sexual contact (e.g. shaking hands or hugging) and insect bites. Epidemiological data have continued largely to characterize HIV/AIDS as critical for men who have sex with men and intravenous/injecting drug users (IDUs); over eight in ten of the cumulative reported AIDS cases nationwide by 1996 have occurred among these two groups (Centers for Disease Control and Prevention 1996; Table 1.2).

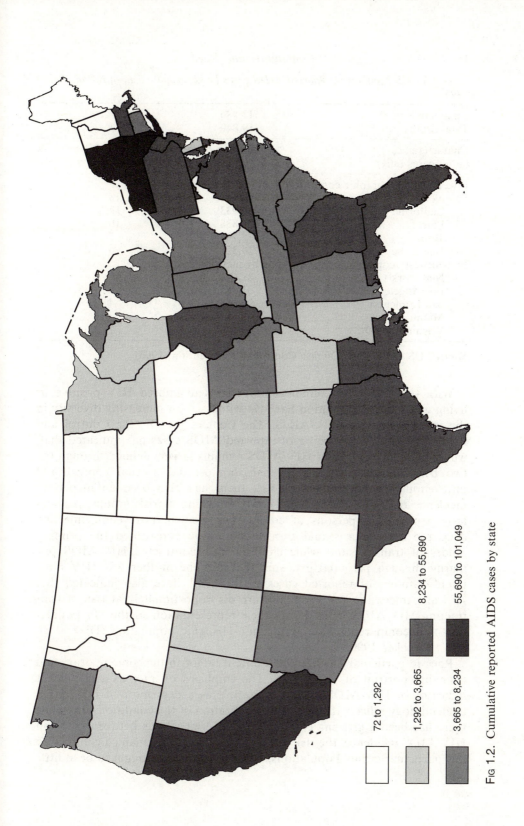

FIG 1.2. Cumulative reported AIDS cases by state

72 to 1,292

1,292 to 3,665

3,665 to 8,234

8,234 to 55,690

55,690 to 101,049

TABLE 1.1. *US Cumulative reported AIDS cases by geographic concentration, June 1996*

Cumulative cases	513,486	
Total deaths	319,849	
Characteristics (adult cases only)	Number of cases	% of cases
Five highest geographic concentrations by state		
New York	94,751	18
California	88,933	17
Florida	51,838	10
Texas	35,144	7
New Jersey	29,327	6
Five highest geographic concentrations by metropolitan area		
New York City	81,604	16
Los Angeles	31,085	6
San Francisco	22,835	4
Miami	16,372	3
Washington, DC	14,640	2

Source: US Centers for Disease Control and Prevention (1996)

With the rise in the numbers of persons diagnosed HIV positive or living with AIDS, the nation has also witnessed an increasing diversity in the epidemiology of HIV/AIDS. The Centers for Disease Control and Prevention's (CDC) tracking of reported AIDS cases has indicated that, while the epidemiology of HIV/AIDS remains largely defined through the two behavioural categories of male homosexual/bisexual contact and intravenous/injecting substance use, there has also been an increasing incidence of HIV/AIDS among newly emerging at risk groups, particularly women and persons of colour. For women, intravenous/injecting drug use and heterosexual transmission have constituted the primary modes of transmission, while children have contracted HIV/AIDS primarily through perinatal exposure (because the mother was HIV positive).[15] Studies of reported cases to the CDC have also indicated that African Americans and Hispanics are disproportionately at risk of contracting HIV/AIDS, considering their representation in the US population and compared to the White non-Hispanic population (Diaz *et al.* 1993; Drucker 1990).[16]

Race in particular has become a potent factor in the contemporary and changing portrait of HIV/AIDS. The emphasis on race as a significant indicator of HIV/AIDS transmission and as a descriptor of HIV/AIDS epidemiology reflects the ongoing racialization of the condition. But while there has been a growing emphasis on the connections between race and HIV/AIDS incidence, the risk of HIV/AIDS has not remained confined to any particular group. Populations, for example, once thought to be at little

TABLE 1.2. *US cumulative reported AIDS cases by socio-demographic characteristics, June 1996*

Cumulative cases	513,486	
Total deaths	319,849	
Characteristics (adult cases only)	Number of cases	% of cases
Gender		
Male	434,719	86
Female	71,818	14
Race/Ethnicity		
White	243,107	47
African American	174,715	34
Hispanic	90,031	18
Asian/Pacific Islander	3,555	1
Native American/Alaskan	1,333	0
Unknown or none reported	745	0
Exposure		
Male homosexual/bisexual contact	259,672	51
Injection drug use (IDU)	128,696	25
Male homosexual/bisexual contact and IDU	33,195	6
High risk heterosexual contact	40,037	8
Receipt of blood transfusion	7,433	2
Haemophilia/coagulation disorder	4,107	1
Unknown or none reported	33,397	7

Source: US Centers for Disease Control and Prevention (1996)

risk of contracting HIV/AIDS are now showing increasing risk of HIV infection (Cochran *et al.* 1991). I will return to these issues in Chapter 5.

The definition of HIV/AIDS has been managed primarily through the Centers for Disease Control and Prevention (CDC). This has meant that government funding and programmes for HIV/AIDS treatment and prevention have been available only to those fitting the definitions constructed by the CDC. This has had particular manifestations in the epidemiological portrait of HIV/AIDS, particularly as it relates to women. Since the early data on HIV/AIDS in the 1980s indicated that HIV/AIDS affected men almost exclusively, there was limited acknowledgement of the gynaecological conditions associated with HIV/AIDS in women, and hence women living with HIV/AIDS were likely not to be recognized as such (Anastos and Marte 1991).

Homelessness and HIV/AIDS

Although researchers, policy-makers, human service providers, and the public continue to view and respond to homelessness and HIV/AIDS as distinct issues, there is growing evidence that the two crises have become increasingly interconnected. Homelessness often has associated with it highly risky behaviours when considering the transmission of HIV/AIDS. Research has indicated for example that between 13 per cent and 52 per cent of homeless persons are users of street drugs (intravenous/injecting drug use remains a primary mode of transmission for HIV/AIDS), and sexual assault and sexually transmitted diseases (particularly among women who are homeless) are widespread (Nyamathi 1992). Homeless persons living with HIV/AIDS also tend to be at greater risk of tuberculosis and other diseases than domiciled individuals living with HIV/AIDS (Layton *et al.* 1995; Lebow *et al.* 1995). There are also interactions among race, homelessness, and HIV/AIDS. A study in Boston found that homeless persons living with HIV/AIDS were more likely to be African American or Latino than homed persons living with HIV/AIDS (Lebow *et al.* 1995).

In addition to the marginalization associated with homelessness, diagnosis as 'HIV positive' and its consequent stigmatization further complicates access to the housing and labour markets (Drucker 1991). As such, HIV diagnosis is a contributing factor to incipient homelessness. As Arno (1991) points out, persons living with HIV/AIDS have in the past been evicted (the Americans with Disabilities Act of 1992 has since made this practice illegal), family members have at times been unwilling or unable to care for a person living with HIV/AIDS, and income earning has been made problematic because of unemployment (both because of inability to work and the barriers to employment erected because of societal stigmatization). Even with such increasing evidence of the links between homelessness and HIV/AIDS, the public, service providers, researchers, and policy-makers continue to address them as relatively distinct.[17]

Dealing with HIV/AIDS

Prior to organized governmental involvement in HIV/AIDS, there were significant gains made in responding to HIV/AIDS during the early 1980s by *ad hoc*, community-based, volunteer organizations comprised of persons affected by HIV/AIDS (Altman 1994; MacLachlan 1992). Two community groups in San Francisco for example (Gay Men's Health Crisis and Shanti) both provided home care and support services for individuals living with HIV/AIDS. A primary component of such community-based services was the designation of 'buddies' for those living with HIV/AIDS. Buddies were comprised of volunteers who interacted with persons living with HIV/AIDS in various capacities, helping with cleaning house, shopping, providing support and conversation, and, in general, providing a different facet of support from that which could be provided by health care workers, friends, family, or lovers/spouses (Altman 1994).

Public health efforts to curb HIV transmission during the 1980s evolved in *ad hoc* and haphazard ways as community-based organizations and government agencies at the municipal, state, and federal levels struggled to respond to the widening crisis of HIV/AIDS. The federal government plays a central role in ongoing efforts to develop a co-ordinated and comprehensive strategy for addressing the changing nature of HIV/AIDS. Federal funding of health services research for HIV/AIDS, for example, rose from $145,000 in 1986 to approximately $43 million in 1991 (Rudzinski *et al.* 1994).

In response to the drastic rise in the number of people living with HIV/AIDS, the US Congress passed the Ryan White Comprehensive AIDS Resources Emergency (CARE) Act in 1990, providing funding for outpatient and ambulatory medical and support services for persons living with HIV/AIDS. This Act, whose funds are administered by the Health Resources and Services Administration (HRSA), was divided into four titles to reflect the diffusion and distribution of persons living with HIV/AIDS and to provide assistance for development and provision of health and support services. Title I of the CARE Act (commonly referred to as Ryan White Title I funds) bases funding for eligible metropolitan areas (EMAs) on reported AIDS cases. Title II supports state-level efforts at improving quality and access of health care and support services. Title III funds early intervention services provided by community health centres and non-profit organizations. Title IV provides funding support for comprehensive service programmes for children, adolescents, and families (Amaro and Hardy-Fanta 1995). Funds were first allocated in fiscal year 1991, when $86 million was allocated to EMAs reporting more than 2,000 cumulative cases to the CDC (Bowen *et al.* 1992). Funding through Ryan White CARE allocations is tied to local level planning processes, requiring that voting members of local planning councils include persons living with HIV/AIDS, health care providers, community-based organizations, public health agencies, non-elected community leaders, and state government officials (Bowen *et al.* 1992).

But federal legislation has also had to deal with the complexities inherent in the political and medical nature of HIV/AIDS. Federal level policy-makers, for example, have not generally supported needle-exchange programmes as a method of controlling the spread of HIV among intravenous/injecting drug users even though case studies have indicated that such programmes lead to both a reduction in HIV transmission risks and the participation of drug users in treatment programmes (Ginzburg 1993; Kaplan 1993). Such programmes are perceived by policy-makers as coming into direct conflict with the 'war on drugs'.

The changing epidemiology of HIV/AIDS in the US indicates that HIV/AIDS has become critical for greater numbers and differing groups across the nation. Although medical researchers continue to search for a cure for

HIV/AIDS, its continuing and widening impact has compelled public policymakers to focus on curbing transmission through education and prevention policies.[18] Education and prevention have also comprised a mainstay of public policy measures addressing HIV/AIDS because of the escalating cost of treatment; the lifetime cost of treating a person with HIV from the time of infection until death has been approximated at $119,000 (Hellinger 1993). One of the most visible education and prevention strategies was an information and education campaign launched in September 1987 by the US Department of Health and Human Services and the Centers for Disease Control and Prevention (CDC). This campaign, named 'America Responds to AIDS,' comprised television and radio announcements, printed materials (posters, advertising, and transit cards), a national AIDS telephone hotline, a national AIDS-information clearing-house, and technical assistance programmes (Bush and Boller 1991).

Although the CDC estimated that 80 per cent of all US households were exposed to the 1987 education campaign (CDC 1988), the federal government did not rely solely on mass media information dissemination in its HIV/AIDS education and prevention strategies. Instead, the CDC's HIV education and prevention strategy, beyond public information, also incorporated education for school-aged populations, individual counselling and HIV testing, and risk reduction information (Roper 1991).

The CDC has also worked to develop community-based models of education and prevention in its effort to contain the spread of HIV/AIDS. Based on the community participatory model developed in the Ryan White CARE Act, in 1993 the CDC instituted a new community planning process as a requirement for HIV/AIDS education and prevention funds (Takahashi and Smutny forthcoming). This new planning process was initiated in response to local concerns about the complexity of existing funding mechanisms and the multiple funding streams created during the 1980s and early 1990s. The HIV Prevention Community Planning Process requires that local knowledge and epidemiologic data be used as the basis for developing and funding programmes. Across the nation, state and local health jurisdictions are developing institutions to incorporate this new requirement for community participation. In California, fifty-four Local Planning Groups (LPGs) worked to develop three year prevention and funding plans by December 1995 based on local epidemiologic portraits, assessments of community resources for HIV/AIDS prevention and education, and the participation of groups affected by HIV/AIDS (persons living with HIV/AIDS and groups identified as being at high risk). I will return to the issue of intergovernmental relations, community participation, and HIV/AIDS education and prevention in Chapter 9.

As the medical community, policy-makers, and service providers have begun to realize that persons diagnosed with HIV/AIDS may live a relatively long and active life (i.e. that HIV diagnosis does not mean an immediate death or even a steady and downward decline), there has been the growing concern that quality of life for those living with HIV/AIDS be enhanced. Such concerns may also have been spurred by persons with HIV/AIDS and concerned groups, which have increasingly become politically mobilized over the issue of HIV/AIDS. Such mobilization has worked towards multiple goals: for public recognition, to increase access to medical research and treatment, and to improve societal response to persons living with HIV/AIDS. Such groups have emerged largely from the gay/lesbian/bisexual communities (Shaw 1991).

Geographers have been at the forefront of constructing theorizations and portraits of the socio-spatial implications of changing quality of life and increasing mobilization. Wilton's (1996) elaboration of life paths and routines after HIV diagnosis provides a critique of the assumption still widely held by the public that being HIV positive necessarily means a steady downward physical and mental decline until death. His ethnographic research with HIV-positive men living in Los Angeles indicates that while HIV diagnosis does bring with it many constraints on the everyday practices of individuals (because of physical and social obstacles), HIV diagnosis can also be dealt with in supportive ways. Geltmaker (1992) outlines the ways in which ACT-UP (AIDS Coalition To Unleash Power), an AIDS activist organization, has worked to alter the spatial politics of HIV/AIDS in Los Angeles. There have also been calls by anthropologists and other social scientists to integrate the voice of persons living with HIV/AIDS into research and public policy (MacLachlan 1992).

But while many persons living with HIV/AIDS, activists, researchers, and service providers have worked to change the public conceptualization of HIV/AIDS, public reaction to HIV/AIDS has tended to remain rejecting. Such rejecting tendencies have manifested themselves through community resistance to individuals living with HIV/AIDS in public places, such as public schools (the case of Ryan White is probably the most well known), and to human service facilities associated with HIV/AIDS in residential communities.[19]

Conclusion

In the United States, the rise in homelessness and HIV/AIDS is at its core geographical in nature. Individuals become homeless as the result of local housing markets, welfare state practices, and the changing upsurges and declines in the labour market. The individual experience with homelessness (whether by those homeless or those who come into contact with homeless persons) is at the local level. Individual homeless persons come into contact with regional, state, and federal government agencies through their

attempts to acquire housing and human services, and for accessing monetary benefits. Homeless populations are also managed at the local level
through municipal ordinances and land-use regulations which monitor and
control where homeless services and shelters (and persons) should be
located. The overall national portrait then is a mosaic of these local
experiences, as well as a national crisis being dealt with by various levels
of government.

    HIV/AIDS is both distinct and linked with the societal crisis of home-
lessness. HIV/AIDS is intricately linked with homelessness in at least two
ways. Those living with HIV/AIDS are often driven to the brink of poverty
because of obstacles to employment, health care, and housing. And those
homeless are at great risk of becoming HIV positive, through the same
modes of transmission affecting the larger population. However, HIV/
AIDS continues to be seen as distinct from homelessness by policy-makers
and the public. Unlike homelessness, HIV/AIDS has been defined as a
disease, with its definition (and concomitant resource distribution)
controlled by the federal level Centers for Disease Control and Prevention
(CDC). As a disease, HIV/AIDS has remained largely in the realm of
biomedical researchers and public health officials. But while the management of the definition and policies responding to HIV/AIDS has largely
been centralized at the federal level, the public response to HIV/AIDS has
been differentiated at the local level (where the public responds in contextually specific ways to individuals living with HIV/AIDS and facilities
providing services to such individuals). To understand the changing
character of both homelessness and HIV/AIDS, we must explore not
only the shifts in size and composition of these two populations and the
changing recognition of homelessness and HIV/AIDS as social and
physical conditions, but also how the public has responded to these crises,
both in trying to ameliorate homelessness and HIV/AIDS, but, more
particularly, with respect to the NIMBY syndrome, in opposing and
removing their presence from neighbourhoods and regions.

2

Explaining Community Opposition

THE print media is replete with examples of residents and municipalities acting to prevent the potential siting of homeless shelters, group homes, and other facilities which might serve homeless persons or people living with HIV/AIDS, or to close facilities which currently exist. A recent national study conducted by the National Law Center on Homelessness and Poverty found for example that among 61 cases of community opposition towards facilities associated with homelessness, 21 resulted in halting projects being designed or implemented, 6 resulted in existing facilities being forced to close, and 2 resulted in coerced moves (cited in Anonymous 1996). The rising tide of rejection at the local level, and the public opinion it represents, constitute significant reasons for the lack of political will often cited as an obstacle to implementing policy recommendations which might address the structural causes of homelessness and HIV/AIDS (e.g. Dear and Wolch 1987; Kress 1994; Patton 1990). This rising tide of rejection is also associated with the upsurge in political conservatism experienced over the past few decades. Such conservatism has recently become intensified, resulting in renewed attacks on the welfare state and individuals using human services. Welfare programmes have been increasingly targeted for reduction or elimination, particularly for specific populations thought to be undeserving or ineligible for state-provided benefits.

As the recent intensification in the US of attacks against the welfare state indicates, local rejection and opposition directed towards homeless persons, people living with HIV/AIDS, and facilities serving such populations are not monolithic across time and space. To understand the socio-spatial variation in response to homelessness, geographers and other social scientists have investigated the processes leading to the rejection of facilities and individuals. There are two primary strands in geographical and planning research on the NIMBY syndrome. The first posits the centrality of production relations in creating undesired outcomes and noxious land uses and facilities, indicating that perceptions of threat are embedded in capitalist social relations. This perspective indicates that responding to the NIMBY syndrome will never be as simple as changing the perceptions and attitudes of individuals and communities, rather, that the fundamental sources of practices such as community opposition lie in structural forces and institutional relationships. The second strand of research centres on the community's perception of the threat of the facility and its clients (Hogan 1986). Following this line of reasoning, the perception and definition of threat are important for understanding why specific communities oppose facilities and why others do not. In this chapter, I present these two major strands of scholarship to contextualize the ongoing academic exploration of the NIMBY syndrome.

The Social Relations of NIMBY

Peter Hall in 1989 described the NIMBY syndrome as 'the populist political philosophy of the 1980s' (p. 280). As an expression of populist politics, or the politics of ordinary people, the NIMBY syndrome, echoing the rise in New Right politics since the early 1970s, comprises efforts by individuals and groups wanting to maintain order and the *status quo* (following Boyte *et al.* 1986). But further, Cornel West (1986) explains that, at its core, populist politics (such as the NIMBY syndrome) operate from two basic premisses, 'relative cultural homogeneity and persistent economic growth of the US capitalist economy' (p. 209).

As West indicates, one framework for understanding community opposition centres on the search for political and state mechanisms facilitating opposition, and the identification of reasons for vehement response to specific types of land uses and facility types. This body of research, using the lens of production relations, has tended to conceptualize the NIMBY syndrome as class conflict manifested in the maldistribution of undesirable facilities and land uses, and, in addition, has explained the scholarly and policy emphasis on community opposition as overlooking the interrelationships among the state, capital, and community (following Lake 1993). In so doing, these researchers have worked to integrate a conceptualization of community opposition in a larger framework of capitalist social relations, developing explanations of locational conflict which go beyond placing 'the onus for conflict on myopic local communities blocking the unimpeachable objectives of good health and a clean environment for everyone' (Lake and Johns 1990: 489).

Using this perspective, community mobilization and opposition (where community is defined broadly, including residents, developers, and institutions involved in production processes) constitute an inherent component of the relationships among capital, the state, and community. As Sidney Plotkin (1987) has argued about land-use politics, 'expansion is the condition of exclusion just as exclusion is the condition of expansion' (p. 10; also Elkin 1987: ch. 2; Popper 1981). That is, exclusionary politics such as the NIMBY syndrome comprise an integral component of the processes of growth and change. Growth and change inevitably create negative externalities (such as environmental hazards, pollution, and homelessness), which may spark affected communities into action (Lake 1993). The lack of production of affordable housing, for example, is intricately linked to the dynamics of the housing market, especially the trend towards demolition of low-cost, larger units, conversion to luxury apartments and condominiums, and gentrification (Baer 1986*a*). Thus, as Plotkin (1987) argues, 'Intra-class housing disputes are promoted by the inter-class reality that housing is a commodity produced by capital for profit. In urban land fights, the organizing formula is: No profit, no housing, plenty of conflict' (p. 34).

In general, and often supporting the wants and desires of those with relative power and influence, the solutions chosen to remedy such conflict will enable the continuation of production and development. As Robert Lake (1993) has argued, 'LULUs [Locally Undesirable Land Uses] constitute structurally constrained political solutions to economic problems that privilege the needs of capital' (p. 88). Thus, classes aligned with capital interests tend to benefit from such solutions, leaving those without connections to power and influence having to bear the burden of undesirable, noxious, or dangerous land uses or facilities. And even when working-class neighbourhoods organize to repel undesirable facilities, such community mobilization tends to be localized, focused on the particular problems being experienced in the confines of the neighbourhood. Thus, the potential of the NIMBY syndrome for large-scale, long-term impacts by low income and working-class neighbourhoods on the land-use system is minimal (Castells 1983; Piven and Cloward 1977). As Logan and Molotch (1987) describe, 'Rich neighborhoods . . . are better able to protect themselves through "working within the system"; poor neighborhoods are particularly vulnerable to disruptions from the surrounding exchange system' (p. 39). This 'working within the system' has often meant the effective use by middle- and upper-income community residents of the state's urban planning apparatus, especially the mechanism of zoning.

The role of state apparatuses as central to the NIMBY syndrome has encountered limited scholarly attention. Indeed, Robert Lake and his co-authors have worked to refocus scholarly and policy debates concerning community opposition and the NIMBY syndrome from solely the outcomes of facility siting (including concerns over risk and equity) to a broader view encompassing the regulation of hazardous waste and other undesirable but inevitable outcomes of production and development. Lake and Disch (1992) argue that the 'challenge to the state is to devise a strategy for waste regulation that fends off the impending legitimation crisis [inherent in the state's relationship with capital] while allowing production (accumulation) to continue virtually unabated. . . . The siting strategy serves to deflect political conflict away from a potentially daunting

challenge to the state–capital relation and into a debate over location'
(p. 665).

The state has various mechanisms available to it to implement such
regulatory strategies with the imagined and material goals of protecting
public health and encouraging development and production. One of the
most ubiquitous mechanisms consists of the design and implementation of
zoning ordinances and regulations. Zoning has been widely effective in
defining and enforcing spatial boundaries between desirable and undesirable land uses and populations, and, writes Sidney Plotkin (1987), 'is best
understood as the protective public armor of landed property in a nation
that holds trespassers in contempt' (p. 22). Zoning has constituted a typical
means of preventing the siting of group care and other residential facility
types especially since the 1960s and early 1970s, when deinstitutionalization and community-based care were pursued by mental health care
professionals and public policy-makers (Lauber with Banks 1974).[1]
Because residents in group homes and other residential facilities do not
fulfill the definition of 'family' in districts zoned for single family dwellings,
human service providers are compelled to obtain zoning variances or
conditional use permits for lawful siting thereby allowing residents to voice
their opposition. This mechanism has proved very successful for preventing
the siting of facilities and land uses deemed undesirable by residents; as
Lauber (1973) comments about multifamily housing, 'Where apartments
are possible only as special uses or variances, the request for such a change
is, and has been, easily denied' (p. 3). Siting human service facilities has
remained contentious, even with US Supreme Court decisions invalidating
zoning ordinances directed specifically at human service facilities (e.g. *City
of Cleburne* v. *Cleburne Living Center* in 1985) and federal legislation
explicitly making illegal discrimination in housing and employment based
on mental or physical disability (e.g. Americans with Disabilities Act of
1992).

Many scholars have argued that through the intersection of multiple
processes (including the urban renewal and revitalization programmes in
downtowns, changing welfare state policies, municipal restrictions and
zoning ordinances, effective opposition in urban and suburban neighbourhoods, and a lack of opposition in Skid Row and older urban areas),
human service facilities have become highly concentrated in 'service-
dependent ghettos' in Skid Row areas near downtowns (Lauber 1985;
Wolch and Gabriel 1985; Wolpert and Wolpert 1976). As Wolch and
Dear (1993) describe the distribution of human services associated with
homelessness in Southern California,

The distribution of homeless-shelter/service programs was strongly linked to
economic class. Gatekeeper (especially shelter) programs were heavily concentrated
in working- and poverty-class communities, which had over half the shelter

programs. Middle- and upper-class communities offered few gatekeeper programs and had a negligible share of shelter programs but much larger shares of coping and other human service programs. (p. 169)

Local governing bodies and service providers often act to avoid the usually confrontational debates encountered in the siting of facilities associated with homelessness and HIV/AIDS. Cities, for example, will use isolationist (e.g. unpopulated locations), circumventional (e.g. not discussing plans for siting), or co-operational (e.g. negotiation) strategies for siting controversial human services (Gaber 1996). Such strategies, however, tend to exacerbate already inequitable distributions of human services. In addition, cities have pursued increasingly punitive policies for excluding and criminalizing practices associated with homelessness, such as municipal codes making sleeping outdoors illegal and police sweeps of homeless encampments (Blau 1992; Wolch and Dear 1993).

Exploring Attitudes and Behaviour

The second logic used by geographers, planners, and other urban scholars has centred on attitude-behaviour models to explain opposition to controversial human service facilities. Attitude-behaviour models focus on perceptions and evaluations of perceived risk and threat, and then predict behaviour based on assumptions concerning either the minimizing of this perceived threat or the compensation for exposure to these risks. Using such models, researchers have conceptualized opposition to human service facilities and their clients as the endpoint in a process beginning with individual characteristics and beliefs, and ending with behavioural intentions and behaviour.

The foundational work in geography using this framework to explore human service facility siting and community rejection sought the linkages between socio-spatial and attitudinal attributes, and rejecting and accepting behaviour (Dear and Taylor 1982; Smith and Hanham 1981). Significant findings from this research (which build upon public facility location models and incorporate social-psychological approaches concerning perception and behaviour) indicated that specific types of residents (such as suburban residents and home-owners) tended to reject human service facilities and clients, that facility characteristics significantly influenced the perception of risk to the community, and that proximity was a primary determinant of the degree to which residents rejected or accepted facilities (Green *et al.* 1987; Segal *et al.* 1980).

Since the early 1980s, geographers have investigated the constitution of the perceived threat posed by human service clients and facilities to communities and businesses, and have developed a four-component typology illustrating the elements associated with threat (Smith 1981; Dear and Taylor 1982). This typology consists of: the characteristics of the client population; the characteristics of the potential host community; programmatic and physical characteristics of the facilities; and the proximity of the facility to residents and businesses in the neighbourhood (Table 2.1). The significance of various characteristics of the client populations and potential host communities on attitude formation is drawn from the premiss that differences put 'social distance' between people, which create concerns and fears about the potential dangers posed by clients and facilities.

The characteristics of the clients who will use a facility has a significant effect on community response, in particular, their demographic attributes, the type of stigma and danger associated with various disabilities, and the visibility of disabilities or behaviours (Takahashi 1996). These socio-demographic and behavioural characteristics of potential clients of facilities translate into assessments about the probable threat posed by proximate clients to life and property. The combination of these client characteristics results in the public's development of hierarchies of acceptance, or the ranking of facilities and clients by their relative degree of acceptability (Dear 1992). The ranking of clients and facilities is based on an assessment of the sum of perceived client characteristics, with the criteria for assessment reflecting the normative community values subscribed to by residents. That is, clients tend to be less acceptable the more demographically distinct they are from the community, the more stigmatized and dangerous they are perceived to be, and the greater the attention generated by their physical appearance and behaviour (Dear and Gleeson 1991). These assessments of acceptability and non-acceptability stem from the degree to which each client group fulfils the normative community values held by residents.

Different communities will react in varying ways to individuals and facilities associated with homelessness and HIV/AIDS based on neighbourhood demographic, locational, and attitudinal characteristics. Geographers have studied such characteristics in the search for connections between neighbourhood characteristics and community attitudes towards human services and clients. Past research has indicated, for example, that accepting neighbourhoods tend to be characterized by larger proportions of unmarried, younger, and renting individuals (indicating a more transient residential population), while rejecting communities tend to be more homogeneous (Smith and Hanham 1981). Recent anecdotal evidence suggests, however, that rejection is now less tied to socio-economic status and class. For example, many racial minority and low-income communities are becoming increasingly involved in local disputes over facility siting (Arrandale 1993).

Attitudinal characteristics of residents in the potential host community also play a central role in the response of that community to the siting of a

TABLE 2.1. *Typology of dimensions of attitude formation*

Element	Key Dimensions
Characteristics of client population	• Demographics (age, gender, *race, ethnicity,* social class) • Type of disability (physical, mental, social) • Severity of disability (contagious, life threatening, chronic, mild) • Visibility of disability (invisible, predominant) • Culpability (blameless, blameworthy)
Characteristics of potential host community	• Location (metropolitan area, suburban area, city size) • Demographics (social class, home-ownership, education, gender, presence of children, marital status) • Neighbourhood homogeneity (social, economic, political, physical) • Attitudinal type (authoritarian, benevolent, socially restrictive, community mental health ideology)
Programmatic and physical characteristics of facilities	• Strategies used to site facility (low-profile versus high-profile approach, risk aversion, fair-share principles) • Size of facility (physical, number of clients) • Type of facility (residential versus non-residential) • Number of facilities in neighbourhood • Operating procedures (degree of supervision, schedule of activities, opening times, outreach programmes) • Reputation of sponsoring agency • Appearance of facility (maintenance, degree of fit in terms of neighbourhood character, landscaping)
Proximity to residents and businesses	• Resident and business awareness of facility • Geographic distance to residents and businesses

Source: Based on Dear (1992).

human service facility. Dear and Taylor (1982), in their research on community response to mental health care facilities, typologized these attitudinal characteristics into four groups: authoritarianism (i.e. the belief that facility clients are inferior and require coercive supervision); benevolence (i.e. a paternalistic view of clients); social restrictiveness (i.e. the belief

that clients pose a threat to the community); and a community mental
health ideology (i.e. the belief that clients should be cared for in local
communities). Communities exhibiting authoritarian or socially restrictive
attitude characteristics will generally be more rejecting of human service
facilities than those reflecting benevolent or community mental health
ideology attitudes.

When faced with a facility serving homeless persons, residents are often
concerned about the programmatic and physical characteristics of the proposed or existing facility. Programmatic characteristics include the strategy
used to site the facility (i.e. visible and high profile versus more secretive and
low profile), operating times, services provided, and the number of staff on
site. Physical characteristics include the architectural design of the facility,
the existence of off-street parking, and the degree to which the facility is
visible from the street. Such characteristics are important for service
providers since they can be directly manipulated to offset community rejection.[2] Six components of the facility have been found to play an especially
influential role in neighbourhood perceptions (Dear and Taylor 1982): type
of facility (related to the type of client being served); size (the size of the
building and surrounding space); number (the number of clients to be
served and number of staff); operations (the general scale of operations
for the facility, including the level of supervision for clients and the daily
routine of the facility); appearance (the visibility of the building and its
clients to neighbours and passers-by, and the quality of the maintenance of
the facility); and reputation (of the organization which is to operate the
facility, or of other human service facilities in the area).

Proximity to a facility or persons associated with homelessness or HIV/
AIDS is the final component of the typology characterizing perceived
threat to communities. In general, both rejection and acceptance tend to
intensify the closer the facility is sited to residents and businesses in the
neighbourhood (Dear and Taylor 1982; Smith 1981). Taylor and Dear
(1981) concluded that a consistent distance decay function (a steady
decrease in attitude intensity as the facility was located further away
from a residence) existed for the four attitude characteristics discussed
above. For example, in their Toronto neighbourhood study, seven to twelve
blocks distance between a facility and a residence elicited less authoritarian
attitudes than a facility located less than a block away. Proximity also plays
a significant role in the externality effects often feared by residents and
businesses concerning persons and facilities associated with homelessness
or HIV/AIDS. Concerns include the potentially negative fiscal impact on
businesses given an increase in homeless persons or people living with HIV/
AIDS in the vicinity, pecuniary effects (such as decreasing property
values), and the changes in traffic, crime, and other neighbourhood
amenities that the presence of facilities and clients might foster. While
studies have indicated that such facility types do not create these negative

externality effects, the fear of these potential impacts continues to linger (Anonymous 1987).

Proximity may also act as a catalyst for concerns about facility saturation. Such concerns emanate from the perception among residents and businesses that their neighbourhoods have been targeted unfairly for human service facility siting especially in relation to other communities. Neighbourhoods consisting primarily of racial minorities and low-income households have been particularly vulnerable to planning strategies which have targeted specific communities for multiple facilities (Wolpert and Wolpert 1976). Perceptions about the inequitable distribution of undesirable facilities have spurred debates and policies concerning fair share allocation of facilities across municipalities and jurisdictions (Dear and Wolch 1987; Gaber 1996; Rose 1993; Weisberg 1993). The even distribution of facilities across regions might have drawbacks, however. With a spatial concentration of facilities, there are several potential positive benefits for clients and service providers. A 'service hub' of facilities, for example, may provide improved access for clients when compared to a system of geographically distant facilities. Such a concentration of a mix of facilities may eliminate the need for long trips on public transportation systems (Dear *et al*. 1994). In addition, the geographic proximity of facilities means that co-ordination among services might be more feasible given the access that service providers have to clients and other providers. I will return to the issue of fair share in Chapter 10.

This body of research has indicated that perceptions towards varying service-dependent groups change over time. Researchers focusing on perceptions towards clients and facilities have highlighted the relative changes in acceptance and rejection over time by comparing hierarchies of acceptance, which consist of rankings of client groups or facility types according to their relative acceptability in the neighbourhood (Blissland and Manger 1983). Although they constitute highly simplified representations of the dynamic underlying shifts in perception concerning clients and facilities, such rankings do provide a relatively quick method of assessing the degree of change in attitudes over time. A 1970 study in the US, for example, suggested that the least acceptable group was composed of 'alcoholics', 'ex-convicts', and 'mentally ill' individuals (Tringo 1970). People who had physical disabilities or ailments (such as ulcers and asthma) were the most acceptable. A later study in 1980 in Ohio indicated that elderly persons and the physically disabled were much more acceptable than 'mentally ill' persons, 'alcoholics', and 'drug addicts' (Solomon 1983). The variation in sampling and survey methods precludes detailed comparisons, but what is clear is that these hierarchies tend to remain defined over time by the perception of difference from socially established norms. Those persons considered to be the most different from normal are consistently located at the bottom of the hierarchy. There is also evidence of change over time in researchers' definition of controversial client groups. The inclusion of 'drug addicts' in the 1980 study and their absence in the 1970 study indicates the shifting nature of human service client groups over time. In the 1980 study, 'drug addicts' established a new benchmark by which the rest of the client groups could be evaluated.

National Attitudes in the US

Recent national survey data in the US on perceptions of human service facilities, collected in late 1989, indicate that newer facility types have differential effects on the perception of more established and familiar facility types, and that specific facility types remain highly rejected (Takahashi 1997; Takahashi and Dear 1997). The first national survey of attitudes concerning human service facilities was conducted in late 1989 by the Daniel Yankelovich Group (DYG) on behalf of the Robert Wood Johnson Foundation. DYG used a random probability sampling technique to obtain a stratified sample of all adults over 21 years of age in the continental US (N = 1326). A sample of operating telephone exchanges was pre-stratified according to the following criteria (presented in order of stratification): the nine US Census divisions; states within these divisions; metropolitan statistical areas (MSAs) versus non-metropolitan county areas; and the division of locations into cities, towns, and rural areas. A random digit dialling process was used to contact approximately 3,016 persons between 1 and 11 December 1989, and resulted in a 44 per cent response rate. The national sample was relatively homogeneous across various socio-demographic and spatial dimensions. Compared to US Census data, the surveyed population was predominantly White (86 per cent), largely middle and higher income (58 per cent had at least a before tax income of $30,000), older, and was composed of more home-owners.

Eighteen facility types and land uses were included in the survey with fourteen being related to human service provision (listed here in alphabetical order): alcohol rehabilitation facilities, day care centres, drug treatment centres, factories, group homes for mentally disabled persons, group homes for mentally retarded individuals, group homes for persons with AIDS, group homes for persons with depression, homeless shelters, hospitals, independent apartments for mentally disabled persons, landfills, medical clinics treating eyes or allergies, mental health outpatient facilities, nursing homes for elderly persons, prisons, schools, and shopping malls. I focus in this chapter on the fourteen human service facility types. Respondents were asked to assess how accepting they would be of these facility types in their neighbourhoods. Attitudes were measured on a six-point Likert scale, with one meaning 'absolutely would not welcome' a facility in the neighbourhood, and six meaning 'absolutely would welcome' a facility. The meaning of responses between these two extremes was left to the respondents.

The relatively new human service facility types in this 1989 national survey, when compared to earlier attitude studies, included homeless shelters and group homes for persons with AIDS. These recent survey data indicated that the most acceptable facility types included schools, day care centres, nursing homes for elderly persons, medical clinics treating eyes or allergies, and hospitals. Less accepted by the national sample were group homes for mentally retarded persons, alcohol rehabilitation facilities, homeless shelters, and drug treatment centres. The least accepted human service facilities in this survey largely focused on mental disability and AIDS. Group homes for persons with AIDS were the least accepted facility type in this survey. These results indicated that although HIV/AIDS and homelessness have become increasingly important over the past two decades, the facilities addressing these populations were perceived very differently by the respondents. Facilities for persons with AIDS were much less acceptable than facilities providing homeless services.

When attitudes towards the human service facilities associated with homelessness and HIV/AIDS were analysed in greater depth, there were even clearer differences in their acceptability (Fig. 2.1). Approximately 20 per cent of the respondents indicated that they 'absolutely would not

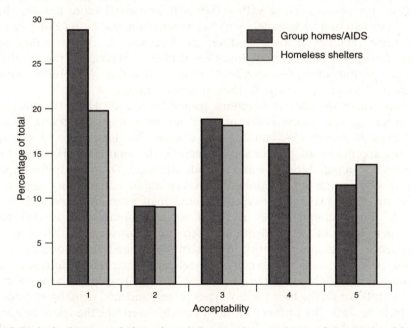

Fɪɢ 2.1. Attitudes towards homeless shelters and group homes for persons with AIDS

welcome' homeless shelters, while approximately 25 per cent reported that they 'absolutely would welcome' these facilities. The remaining 55 per cent of the respondents were somewhat more neutral about this facility type. The facility type associated with AIDS, however, was seen very differently by the national sample. Almost 30 per cent reported that they 'absolutely would not welcome' a group home for persons living with AIDS, while only about 16 per cent reported that they 'absolutely would welcome' this facility type.

This differentiation in attitudes implies that group homes for persons living with AIDS will be viewed more harshly than facilities associated with homelessness. There are fewer respondents who would strongly welcome a group home for persons living with AIDS when compared to homeless shelters. Previous research has indicated that intense negative perceptions are highly likely to translate into rejecting behaviour such as community opposition when a facility is to be sited (Dear and Taylor 1982). Thus, one might expect that group homes for persons living with AIDS might experience a greater degree of organized community resistance. This does not mean that homeless shelters will not face opposition, however. In most cases of facility siting, proposed facilities tend to be viewed singularly rather than in comparison to other types of facilities. It is not a common practice to site multiple facilities at the same time in the same community. Thus, although this figure does suggest that group homes for persons living with AIDS will be viewed more harshly than facilities catering to homeless persons, residents might be rejecting of any of these facilities if sited in their neighbourhoods. What makes this occurrence less likely for the homeless shelter, however, are the relatively larger proportions of respondents stating that they 'absolutely would welcome' this facility type in their neighbourhoods.

But where do these welcoming respondents live? How significant is location in the acceptability of facilities providing services to highly stigmatized groups? We can begin to explore the significance of location in the acceptance of human service facilities by analysing the connection between residential location and attitude structure. To ascertain the importance of location in explaining the variation in acceptability responses towards these two facility types, ordered logit estimation methods were used.[3] The national survey included socio-economic and demographic characteristics of respondents, and spatial characteristics of place of residence. Pearson correlation coefficients indicated that none of these independent variables was highly correlated. These characteristics were used to explore the distinctive role that location played in the variability in acceptance to the four facility types in respondents' neighbourhoods.

In these data, the primary location variable used was the telephone area code where the respondent lived. To compare attitudes across regions, the respondents were grouped according to telephone area code. Telephone area codes provided a finer disaggregation than state boundaries and were the finest scale of disaggregation possible using these survey data. Data in jurisdictions with fewer than three respondents were not reported.

For homeless shelters, the area code in which the respondent lived did not significantly explain the variation in attitudes (Table 2.2). Rather, the only locational variable which was important was whether the respondent lived in a large city or not (i.e. with population over 50,000). For homeless shelters, living in a larger city was associated with less accepting attitudes. One explanation for this finding might be the relative concentration of facilities and facility users in large metropolitan areas. The historical development of Skid Rows has seen not only a concentration of facilities, but also a concomitant concentration of homeless persons. Thus, the

TABLE 2.2. *Ordered logit model of attitudes towards homeless shelters*

Variable	Coefficient	Standard error	z	p< | z |
Area code	−0.0004	0.0003	−1.119	0.263
Large city or not**	−0.4853	0.1754	−2.767	0.006
Metropolitan area or not	0.2600	0.1884	1.379	0.168
Children younger than 6 years*	0.3169	0.1904	1.664	0.096
Children between 6 and 12 years	0.0868	0.1791	0.468	0.640
Children between 12 and 18 years	0.0490	0.1981	0.248	0.805
Children older than 18 years	0.3165	0.2264	1.398	0.162
Older parents in household	0.0782	0.2632	0.297	0.766
Educational level	−0.0701	0.0538	−1.302	0.193
Employment status	−0.0998	0.0810	−1.232	0.218
Number of earners	0.0628	0.0949	0.662	0.508
Own home or not***	−0.6841	0.2114	−3.235	0.001
Length of time lived in neighbourhood	0.0705	0.0636	1.108	0.268
Household income less than $30,000*	0.4072	0.1866	2.183	0.029
White or not	−0.3994	0.2833	−1.410	0.159
Married or not	−0.1590	0.2142	−0.733	0.463
Female*	0.3444	0.1569	2.195	0.028
Hispanic or not	0.1059	0.1734	0.611	0.541

*=p<0.1,   **=p<0.01,   ***=p<0.001

Model Specifications
Number of observations	553
Log likelihood	−935.55
Chi-squared (df = 18)	64.17
prob > chi-squared	0.000

potential or actual siting of homeless shelters might constitute very immediate and material threats for respondents living in large cities.

The analysis also indicated that respondents in stable households who own their homes and who constitute 'productive' members of the community (e.g. being employed, having higher incomes, and being married) were more likely to be rejecting of homeless shelters. Home-ownership, for example, was indicative of less acceptance. However, having children in the household seemed to be associated with greater acceptance after controlling for other socio-economic and demographic characteristics. These findings indicated that the presence of children in households did not necessarily result in overall greater rejection of homeless shelters, as past research has argued, and indeed, the presence of children may even be associated with greater acceptance of certain types of facilities (e.g. homeless shelters).

Group homes for persons living with AIDS reflected a different pattern of significant variables than the facility associated with homelessness (Table 2.3). Specifically, the ordered logit model for group homes for persons living with AIDS indicated that location (measured through area code) played a distinctive role in the variation in acceptance shown towards this facility type. Unlike the facility associated with homelessness, the size of the city where the respondent lived was not significant in explaining the variation in acceptance. Like homeless shelters, having children (with the presence of children older than 18 in the household associated with greater acceptance), home-ownership (associated with rejection), being married (associated with rejection), and gender (being female associated with acceptance) were significant.

Because the location variable proved to be significant for group homes for persons living with AIDS, the responses for this facility type were mapped to explore the regional character of acceptance (Fig. 2.2). One might expect acceptance for group homes for persons living with AIDS to be concentrated in places where the politics of HIV/AIDS might be viewed more sympathetically. To some degree, this is true. The coastal section of California (including San Francisco and Los Angeles) and many north-eastern states (including parts of Massachusetts, Maine, Connecticut, and parts of New York) tended to reflect relatively accepting attitudes. But, counter to popular understanding about the relative lack of acceptance of difference in the Midwest and Rocky Mountains, there were numerous states in these regions where group homes for persons living with AIDS were relatively acceptable. Respondents in Colorado, parts of Kansas, New Mexico, parts of Oklahoma and Texas, and Utah all indicated acceptance in the upper 25 per cent of all respondents in the sample.

Although there was a wide array of perceptions shown towards the human service facilities included in the survey, these attitudes rarely trans-

TABLE 2.3. *Ordered logit model of attitudes towards group homes for persons with AIDS*

Variable	Coefficient	Standard error	z	p< | z |
Area code*	−0.0006	0.0003	−1.643	0.100
Large city or not	−0.1319	0.1765	−0.747	0.455
Metropolitan area or not	0.1738	0.1886	0.922	0.357
Children younger than 6 years	0.2386	0.1924	1.241	0.215
Children between 6 and 12 years	0.0894	0.1818	0.192	0.623
Children between 12 and 18 years	0.1196	0.1959	0.610	0.542
Children older than 18 years*	0.4576	0.2288	2.000	0.045
Older parents in household	−0.0233	0.2715	−0.086	0.932
Education level	0.0596	0.0538	1.109	0.268
Employment status	−0.0192	0.0811	−0.237	0.813
Number of earners	0.0646	0.0963	0.671	0.502
Own home or not**	−0.5828	0.2062	−2.826	0.005
Length of time lived in neighbourhood	0.0857	0.0650	1.320	0.187
Household income less than $30,000	−0.0082	0.1854	−0.044	0.965
White or not	0.1974	0.2702	0.931	0.465
Married or not**	−0.5462	0.2121	−2.575	0.010
Female*	0.2971	0.1572	1.889	0.059
Hispanic or not	−0.0558	0.1747	−0.319	0.749

*=p<0.1,   **=p<0.01,   ***=p<0.001

Model Specifications
Number of observations	554
Log likelihood	−909.49
chi-squared (df = 18)	34.08
prob > chi-squared	0.0123

lated into action. These national survey data indicated that a large proportion of the respondents had not participated in community opposition or support of facilities in their neighbourhoods. Fewer than 10 per cent of the respondents reported participating in either supportive or oppositional activities concerning a proposed facility. This result tends to support earlier studies which suggest that community opposition is composed of relatively small, often well-organized groups. Thus, rather than reflecting the widespread rejection of facilities and clients across a neighbourhood, such researchers have argued that community opposition represents the rejection of a highly vocal minority (Dear and Taylor 1982).

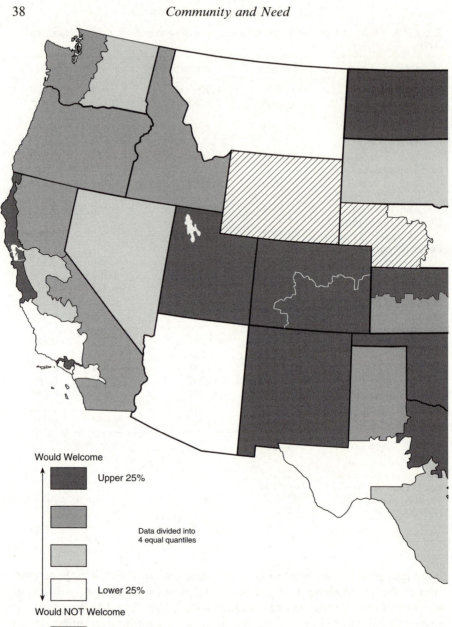

Would Welcome

Upper 25%

Data divided into
4 equal quantiles

Lower 25%

Would NOT Welcome

< 3 Respondents

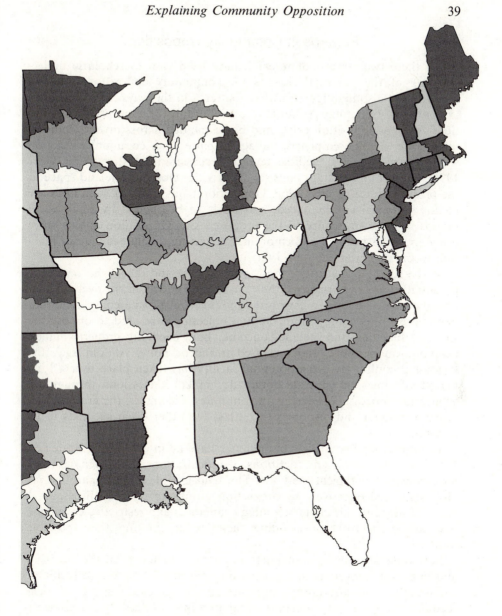

Fig 2.2. Attitudes towards group homes for persons with AIDS by telephone area code

Patterns in Community Opposition

While these two strands of research have been vital to scholarly understanding of the potential reasons for community opposition, ongoing
research has also highlighted various patterns in the ways that opposition
is conducted over time. Although the two strands of conceptual explanations for why opposition exists and has expanded across diverse communities differ in their emphasis, the actual patterns of community reaction
when residents oppose facilities tend to occur in a three-stage cycle (Dear
1976). The 'youthful' stage consists of activities close in time and space to
the proposal for a facility siting. Opposition tends to occur soon after the
facility proposal is announced, and tends to be elicited by small vocal
groups in close proximity to the proposed site. Opposition at this stage
often appears irrational and emotional, reflecting the immediate raw reaction to the potential siting of a facility.

At the stage of 'maturity', those who support and those who reject the
potential siting have organized their members. The arguments over whether
or not to site the facility have often moved from private complaints to a
public arena. Seemingly more rational objections are used to oppose the
siting of facilities, and include concerns about declining property values,
neighbourhood safety, and increased traffic or noise. At 'old age', the
locational conflict over the potential facility has taken place over a long
period of time. In general, both sides make concessions through an
arbitration process. However, with continuing stalemates, the group which
has the stamina and resources to outlast the other often succeeds in its
mission.

To oppose facility siting, communities have adopted various strategies.
Letter-writing campaigns, petitions, demonstrations, and even violence
have constituted typical methods. The land-use planning apparatus has
also been used effectively by opposition groups to reject facilities. As I
have discussed in this chapter, zoning ordinances and restrictive covenants
have been effective in excluding facilities or limiting their possible
locations.

But while scholars and planners often tend to focus on the rejecting
and opposing behaviour of residents directed at human service facilities,
communities are not always opposed to the potential siting of such
facilities. Support may emanate from multiple sources, but arguments
for supporting the siting of such facilities usually revolve around four
themes: the benefits provided by already-sited and operating facilities, the
need for such facilities, legal issues, and calls for charitable or
humanitarian responses to need. Existing facilities are often used as
examples of the beneficial effects of siting and can also be used to
demonstrate the daily operation of proposed facilities. Such examples
can reduce the concerns of the community by demystifying the physical

and operational characteristics of the potential facility (Takahashi 1998*b*). In addition, fears concerning falling property values and increasing crime can be shown through such examples to be unfounded. The unabated need for human service facilities, particularly for those homeless and persons living with HIV/AIDS, is another common argument for siting facilities. The lack of programmes to serve potential clients, the inadequacy of existing services, and the need for communities to take responsibility for delivering such services are three points used to support this argument.

Legal arguments for siting human services in communities have focused on constructing 'as of right' zoning ordinances, the illegality of discrimination against physically and mentally disabled persons, and specific legal cases. Zoning ordinances supportive of facility siting allow such siting without requiring a specific use permit or the approval of the local community. Such ordinances can be described as siting facilities 'as of right', and have proven to be necessary in many communities where prejudice and discrimination have played central roles in the rejection of human service facilities. But siting facilities 'as of right' provides no guarantee that a community will be receptive or accepting of the facility or its clients. Federal legislation such as the Fair Housing Amendments Act of 1988 and the more widely known Americans with Disabilities Act of 1992 (ADA) have made discrimination illegal against mentally and physically disabled persons in housing or employment (I will return to these issues in Chapter 10). Legal strategies have often been used by advocacy groups to mitigate community opposition (Rocha and Dear 1989).

The least used set of supportive arguments concerns calls for charitable or humanitarian responses to the needs for human services. Two points are made: the plight of the client population should be seen as a societal crisis rather than a public nuisance, and the needs of these populations are created by large structural changes rather than being caused by individuals.

Conclusion

Attitude-behaviour studies have proved very useful for policy-makers and service providers in trying to predict what types of communities might be more likely to respond with opposition, and to what types of facilities. Survey data have been particularly useful in tracking changes in the dynamics of perception concerning human services and their clients. These types of attitude-behaviour studies, of which the analysis of the national survey data was illustrative, were ground-breaking in that they worked to clarify and categorize many of the particularities of resident response. Researchers hypothesized that community opposition reflected the reasoned actions of individuals based on their attitudinal predispositions, rather than attributing community opposition to the selfish, irrational response of residents to controversial facilities.

There are several conclusions that can be drawn from the brief analysis of the national attitude survey presented in this chapter. First, the degree of acceptability shown towards the facility types appeared to be only marginally linked with location when measured at the scale of telephone area code. Group homes for persons living with AIDS constituted the only facility type where the area code variable proved to be significant in explaining the variation in attitudes. Second, there was little evidence to support a regional characterization of accepting and rejecting attitudes. From the analysis of the national survey data, acceptance and rejection existed across the US, rather than in strongly defined regions, and therefore, we cannot assume for example that communities in the Midwest might be more rejecting of group homes for persons living with AIDS than communities on the East or West coasts. The third conclusion was that in addition to the large variability in perception within regions, there was also often a large variability in acceptance within state boundaries. This within-state variability might be very much linked to the higher rejection shown by respondents living in larger cities (i.e. those with populations over 50,000).

Using such methods, however, the analysis of such changes remains somewhat descriptive. Survey data may indicate that changes have occurred, and that variation across space does exist, but they are less successful at untangling the reasons for these changes. The reasons for the limitations of survey data, which include but are not limited to surveyor and respondent bias have been widely discussed and comprise substantial sections in survey methods texts (Alreck and Settle 1985).

The perspective of production relations provides alternative explanations for the NIMBY syndrome to the attitude-behaviour logic. Because locational conflict is viewed as an inherent component of production and development (with community opposition occurring in response to the negative externalities inevitably created by such processes), the social relations of the NIMBY syndrome highlight the pervasive nature of community opposition, and indeed of social issues such as homelessness and HIV/AIDS, in contemporary capitalist society. This interpretation of the problematic of the NIMBY syndrome integrates structural relations of the economy, the state, and society into explanations of this phenomenon, and thus provides the opportunity for weaving contextual and localized information about specific cases.

There are, however, substantive reasons to look more broadly than is possible with attitude-behaviour models and explanations centring solely in production relations at the issue of community opposition towards human service facilities. With the increasing size and variation in the service-dependent population, and the growth in local activism concerning many issues, dualistic categorizations may be less useful. Varying types of communities are becoming enmeshed in local opposition (other than the stereotypical suburban, home-owning population). In addition, the size, dispersion, and diversity of the service-dependent population may make the use in surveys of highly charged labels such as 'mentally ill' inadequate indicators of the population's potential response to human services. The issue of labelling and broader issues of representation and identity have proved increasingly salient in ongoing debates on social theory concerning difference. With these conceptual discussions, there are now new avenues with which to discuss and analyse not only the outcomes of the process of community opposition, but also the socially constructed reasons for differentiated rejection and acceptance, and the fragmentation and fragility of such responses over time and space. The variation and changes in the social construction of homelessness and HIV/AIDS over time and space lead to a dynamic of distinction, where differences and abnormality are defined in radically different ways across communities. This dynamic of distinction is reflected in different types of communities opposing facilities, and different types of service-dependent individuals being rejected. It has led to our understanding of community opposition as a complex process reflecting a varied geography of community opposition, locational politics, and urban life.

Part II

Stigmatization and Difference

3

Stigmatization, Homelessness, and HIV/AIDS

Every classification implies a hierarchical order for which neither the tangible world nor our mind gives us the model. We therefore have reason to ask where it was found.

Émile Durkheim and Marcel Mauss (1963)

How might we explain the patterns of public response (both attitudinal and policy) outlined in the previous two chapters? What theoretical frameworks provide a basis for interpreting the rise in localized rejection of human services associated with homelessness and HIV/AIDS? I outlined in the previous two chapters the substantial and growing body of scholarship on homelessness and HIV/AIDS, and on community response to controversial facilities. But as I argued at the end of Chapter 2, community response to homelessness and HIV/AIDS is also based fundamentally in social and spatial relations, particularly of stigma. To provide a theoretical context for a conceptualization of the NIMBY syndrome as socio-spatial stigmatization, this chapter focuses on recent social theory debates concerning difference and classic sociological frameworks defining stigmatization. There are two related issues. One centres on how we should understand the connections between emerging notions of difference and classic definitions of stigma. The second centres on the mechanisms and reasons for the creation and reproduction of stigma around homelessness and HIV/AIDS.

Scholarship on difference has gained popularity in the social sciences particularly over the past two decades. Difference has become a popular framework in social theory for addressing the contemporary dynamic underlying the creation and reproduction of societal and spatial distinction. With rising interest in postmodernist and other poststructuralist frameworks, scholars have explored innovative methods for research and have sought distinctive (and often previously marginalized) and situated perspectives to explain social phenomena (e.g. Dear 1988; Nicholson and Seidman 1995; Rose, G. 1993). But such searching has also been met with amusement, pessimism, and even vehement outrage. Charges have been made about postmodernism's relativism, nihilism, and its destruction of emancipatory projects emanating from modernist perspectives (Davis 1985; Harvey 1989). As critics have pointed out, relativism is not inherently more inclusionary: 'relativism can be a double-edged sword, serving powerful interests above those of the marginalized and dispossessed' (Jackson 1993: 123). Moreover, as David Harvey points out in his discussion of twenty-five deaths in 1991 in a fire at a North Carolina chicken processing plant, 'not only do men and women, Whites and African-Americans die in a preventable event, but we are simultaneously deprived of any normative principles of justice whatsoever by which to condemn or indite the responsible parties' (Harvey 1993: 53). There has also been the argument that essentialism is not at its core necessarily an oppressive approach. As Diane Fuss (1989) argues, 'in and of itself, essentialism is neither good nor bad, progressive nor reactionary, beneficial nor dangerous. The question we should be asking is not "is this text essentialist (and therefore 'bad')?" but rather, "if this text is essentialist, *what motivates its deployment*"' (p. xi, emphasis in text).

No matter where we fall in the debate concerning the relevance of essentialism, this dialogue has made clear the problematic nature of essentialist characterization and categorization. In the context of theorizing community response to homelessness and HIV/AIDS, much of the public and policy response to these two social crises has been framed through the delineation of difference from wider society. Hence, researchers and policy-makers have sought the distinctive nature between homelessness and other forms of poverty, between HIV/AIDS and other terminal illnesses. This approach has fulfilled pragmatic and ideological functions, in terms of defining relatively clear policy objectives (e.g. homelessness is more than house-lessness and therefore requires a continuum of care), and in delineating a relatively clear, understandable identity from which politics can emerge (because of the tenuous and insecure positions indicated by homelessness and HIV/AIDS).[1] However, frameworks which assume essential and natural differences between homeless persons and domiciled individuals, in the longer term, solidify the boundaries between 'us' and 'them', effectively guaranteeing the marginalization of specific groups to less valued social positions and places. Even when studies have incorporated facets of difference, such as the increasing investigation of HIV/AIDS among minorities, they are often incomplete since many begin and end at the description of socio-demographic variations in infection rates, emphasizing HIV/AIDS as 'illness' using public health and medical perspectives (Kearns 1996). Thus, while such perspectives have been useful in framing homelessness and HIV/AIDS for planning and public policy purposes, they have also contributed to the ongoing separation of homeless persons and people living with HIV/AIDS from the rest of 'us'.

Even with calls for nuancing essentialism (e.g. Fuss 1989) and distrust of post-theories (Harvey 1989), the critiques posed by postmodernists, feminists, and scholars of colour of essentialist, modernist frameworks cannot be ignored in the study of public response to homelessness and HIV/AIDS. I argue in this chapter that such theoretical developments are critical to a nuanced understanding of the NIMBY syndrome. Community opposition in specific neighbourhoods towards particular human service facilities are a product of socio-spatial relations such as the stigmatization of difference, with the character of such relations dependent upon particular neighbourhoods' experiences with homelessness and HIV/AIDS. To ignore this issue means the continued *ad hoc* treatment of locational siting conflicts, the implied perception of residents as selfish and exclusionary, and the underlying assumption that somehow homeless persons and people living with HIV/AIDS are not quite as good as the rest of society.

Difference and the NIMBY Syndrome

Difference is used in everyday life to define insiders and outsiders. As Sander Gilman (1985) has argued, 'Difference is that which threatens order and control; it is the polar opposite to our group' (p. 21). When put into practice, difference can be expressed in multiple ways at various spatial scales: through seemingly passive individual actions such as avoidance, and more confrontational individual and institutional actions such as discrimination in employment and housing, or even violence against persons and property.

Particularly over the past decade, debates concerning difference have emphasized the exclusion of persons and their voices by widely accepted frameworks and debates. Such exclusion has dictated which social phenomena and groups retain access to voice, visibility, and power. Through 'violent rhetorical reductions', specific groups have been deliberately and absolutely excluded from discussion to simplify the complexities of everyday life (Ryan 1989; also Deutsche 1990). Such arguments are not new. Feminists have long criticized the disregard of gender issues in planning and social science research (Hawkesworth 1988; Sandercock and Forsyth 1992), and researchers have increasingly problematized notions such as race and ethnicity (Anderson 1991; Hooks 1990; Pulido 1996*a*, *b*).

The power of such critique stems in part from the increasing recognition throughout academe that theory and knowledge cannot be separated from social action. Tactics and strategies used to perpetuate and intensify structures of power are intricately linked to discourse and knowledge (Foucault 1980). This is especially important in considering social phenomena such as the NIMBY syndrome. The theories that people use to describe, explain, and predict social events, such as homelessness and HIV/AIDS, reflect their perceptions of society and consequently their expectations for the future (Toulmin 1989). Conceptualizations based in difference thus become a critical component in our understanding of the pervasiveness of localized rejection shown towards homelessness and HIV/AIDS.

In terms of the NIMBY syndrome, geographers and planners have often sought explanations for localized rejection of human services in theories of social distance. As noted in the previous chapter, while such perspectives have been extremely useful in describing and categorizing the varied nature of communities in their acceptance and rejection of specific service-dependent groups, arguments based in difference have argued that also required is the understanding of the social construction of such distinctions (Dear *et al.* 1997). That is, with difference, there are now new avenues to discuss and analyse not only the outcomes of the process of community opposition, but also the socially constructed reasons for differentiated rejection and acceptance.

Of particular importance is the growing participation of groups which identify themselves racially in the politics of controversial facility siting and land use. The contestation over social relations defined by race, class, gender, sexuality, and age comprise a central component of scholarly debates concerning difference (Anderson 1991; Blunt and Rose 1994; Jackson 1989; Laws 1994). The intersection of such social categorizations with community opposition has problematized the NIMBY syndrome with respect to human service facilities. Although local mobilization over environmental hazards, particularly as it relates to communities of colour, has generally been viewed in terms of empowerment and from the point of view of new social movements (Pulido 1996*a*), similar claims have not been made with respect to human service facilities. The exclusion of groups which are service dependent through such practices as the NIMBY syndrome are deliberate attempts to retain social distance and maintain boundaries between the acceptable and the unacceptable. Thus, unlike mobilization concerning environmental hazards, community organization in opposition to human service facilities has been generally seen as exclusionary and even selfish. But with the increasing participation of communities of colour and low-income communities in the politics of human service provision, the NIMBY syndrome has taken on new dimensions.

Rather than being a monolithic exclusionary practice, the NIMBY syndrome directed against human service facilities can also be argued to have empowering facets. This is, however, a complex issue, pitting marginalized groups against other marginalized groups. For example, communities of colour which have been marginalized across multiple dimensions (political, economic, and social) are increasingly organizing to oppose facilities for service-dependent individuals and families, who have been subject to similar and different processes of marginalization.[2] These degrees of marginality are often intertwined with social relations of race, gender, and sexuality. For example, the dominant representation of HIV/ AIDS as 'White gay male' has catalysed mobilization within gay, lesbian, and bisexual populations for increased access to treatment and information, leading to the solidification of persons and places associated with HIV/AIDS as part of a 'White gay male' identity. However, at the same time, such representation and mobilization has led to the marginalization of other groups (e.g. women of colour living with HIV/AIDS and gay men of colour), with rejection within communities of colour and lack of recognition of such individuals by medical and public policy institutions.[3] Thus, the NIMBY syndrome directed against human services cannot be deemed simply an exclusionary practice by individuals who are members of some monolithic group in power. Rather, to understand the nuances of community opposition towards human services, the broader social relations underlying the dynamics of distinction must be integrated into conceptualizations of the NIMBY syndrome.

As many scholars exploring race, gender, and sexuality have argued, such relations are socially and culturally constructed, and vary over time and space (Gilman 1985; Nicholson and Seidman 1995; Scott 1972). In developing a framework which explains the dynamic of distinction inherent in the NIMBY syndrome, difference plays a central role bacause of its power in describing and explaining the reasons why people might want social distance or coercive treatment of specific individuals and groups (Jackson 1991: 201). Belief systems and practices not only illuminate the dynamic underlying public response to human service facilities, but also have severe material impacts on service-dependent people's capacity for quality of life and even survival (following Pulido 1996*a*; Jackson 1987). One vital mechanism and process which defines difference in terms of community opposition to homelessness and HIV/AIDS is stigmatization, and the role that stigma plays in the development and reproduction of devalued social categories.[4] As a fundamental component of the contemporary dynamic of distinction, this chapter discusses the ways that stigmatization, like other social relations, is 'not natural but subject to invention and change' (Young 1990: 92).

What is Stigma?

The classic sociological notion of stigma describes those processes of social relations that lead to and reproduce definitions of outsider and Other, which, in turn, result in processes of stigmatization which define specific groups as being undesirable, dangerous, unsettling, or disturbing. These frameworks, popularized by Erving Goffman in his seminal work (1963), argue that a process of stigmatization results in individuals and groups becoming devalued in society. Because of some marked difference or 'socially disqualifying attributes', individuals acquire a spoilt or tainted identity (Jones *et al.* 1984; Katz 1981). These differences or socially disqualifying attributes are signs that there are significant discrepancies between society's ideal and homeless persons or people living with HIV/ AIDS, and that these discrepancies constitute extremely negative qualities in persons who are homeless or living with HIV/AIDS. These attributes through time become associated with shame, disgrace, infamy, degeneration, or failure, with the visible nature of the attribute eventually becoming less important than the essence of the difference from socially established norms (Foucault 1973; Gilman 1988; Goffman 1963). The social construction of this discrediting process becomes invisible to wider society, thus responsibility for these tainted identities becomes placed on the individual (Ainlay and Crosby 1986). Placing responsibility for devaluation on the individual reinforces the legitimacy and validity of stigma. Thus, through our collective sharing of the meaning of stigma and deviance, the 'grand illusion' of social order is maintained (Jones *et al.* 1984: 84; also Berger and Luckmann 1966). Or as Edward Soja (1996: 87) has argued about the 'hegemonic power' represented through the simplification of cultural politics of identity and difference into 'hegemonic and counter-hegemonic categories':

Hegemonic power, wielded by those in positions of authority, does not merely manipulate naively given differences between individuals and social groups, it actively *produces and reproduces difference* as a key strategy to create and maintain modes of social and spatial division that are advantageous to its continued empowerment and authority. (emphasis in text)

In general, we can think about stigmatization as comprising systemic and systematic social rejection acting to create and maintain boundaries. Such boundaries help to define 'us' by constructing individuals and groups which are 'abnormal'. Thus, the delineation of deviance and its subsequent and ongoing stigmatization essentializes the differences between wider society and individuals associated with homelessness and HIV/AIDS. Through stigmatization, homeless persons and people living with HIV/ AIDS become defined as essentially different from the rest of the population. This definition as essentially different becomes a socially constructed, designated, and shared negative evaluation of homeless persons and people living with HIV/AIDS (Ainlay and Crosby 1986).

The social and moral limits reinforced by the identification and stigmatization of specific groups promotes the dominant standards in society for appearance and interaction. These limits are implemented through various types of containment strategies (Schur 1980): moral containment (e.g. avoidance), economic containment (e.g. limited access to the labour market), geographic containment (e.g. service dependent ghettos such as Skid Rows), pharmacological and electronic containment (e.g. psychotropic drugs or electronic surveillance), and physical and social psychological containment (e.g. incarceration).[5] The normal/deviant dualism, supported by the stigmatization of difference and the multitude of available containment strategies, reflects the societal process of normalization which brings to society's normal members feelings of predictability, stability, and safety. Stigma allows normal individuals to cope with fearful or threatening conditions by depersonalizing individuals and treating them as 'mere instances of a discreditable category' (Schur 1980: 4). The reason for this fear and anxiety, argue Jones *et al.* (1984), stems in part 'from the fact that such conditions stand as stark reminders of the very things we devote so much of our individual and collective energies to shutting out, ignoring, and avoiding' (p. 86). In so doing, the identification of deviance and difference helps normal society understand and construct its identity; the normal becomes the 'unstated point of reference [used] when assessing others' (Minow 1990: 5).

Not only are individuals able to cope with their fear or perception of threat through the stigmatization of homelessness and HIV/AIDS, but, further, stigmatization enables the discipline of its nonconforming members for maintaining social order. Such discipline occurs often through 'small acts of cunning endowed with a great power of diffusion, subtle arrangements, apparently innocent' (Foucault 1977: 139). Through the delineation of stigma and the discipline of deviance, individuals, groups, and institutions create and maintain 'moral containment' (Schur 1980) of the undesirable, the disturbing, and the dangerous through practices deemed acceptable, and even usual or commonsense (also Douglas 1966; Epstein 1995). So, when we are driving in our cars, and we see a homeless person on the side of the road at a stop light waiting with a sign attempting to elicit monetary contributions from passers-by, we lock our doors and roll up our car windows to shut out the possibility of interaction. The apparently innocent motivation of this practice (i.e. to prevent the predicted and unwanted asking for money) is, as the quote by Michel Foucault indicated, a 'small' act imbued with subtlety and deliberate design.

In exploring stigma as nonconformance to widely accepted social norms, many sociological studies have focused on deviance as rule-violating behaviours (e.g. Lemert 1972; Scott and Douglas 1972).[6] Becker (1963), for example, defined the construction of stigma in this manner: 'social groups create deviance by making rules whose infraction constitutes deviance, and by applying these rules to particular people and labelling them as outsiders.' Frequently studied groups deemed to be rule-breakers include juvenile delinquents, criminals and prisoners, homosexuals and prostitutes, people who murder or commit suicide, substance abusers (alcohol and intravenous drugs), and mentally disabled persons (e.g. Rushing 1969; Spitzer and Denzin 1968; Traub and Little 1985). Such research sought to clarify and explore the norm structures of deviant subcultures to which stigmatized persons were argued to belong. Membership in such subcultures emerged from the inability or lack of desire on the part of deviant individuals to function within dominant norm structures or as an outcome of opposition to dominant group morals and rules.

But stigma can also be formulated more broadly, as ideology. Stigma as ideology means that the rejection of homelessness and HIV/AIDS is not solely the prejudiced beliefs and actions of intolerant individuals, but is based in structural and institutional forces. In this sense, stigma is not monolithic, but can occur among individuals, groups, and institutions in various ways. This notion of stigma as ideology also implies that stigma is political. To understand the politics of stigma requires an exploration of how practices, beliefs, and institutions contribute to marginalization or to collective change. Politicizing stigma means explicitly reflecting on its forms of meanings, embodiment, and emplacement (Laws 1995).

How is Stigma Produced?

Past research has indicated that stigma produces marginalized and privileged identities (Gupta and Ferguson 1992; Schur 1980), maintains existing structures of domination and social control (following Omi and Winant 1994), and facilitates normalization (Foucault 1977). But how is stigma produced?[7] One primary mechanism through which stigma is produced is the process of labelling. In his classic formulation, Lemert (1951) categorized 'deviants' as primary and secondary, with primary deviants behaving in abnormal ways, and secondary deviants being labelled as 'deviant'. The labelling act devalues the individual, effectively separating them from mainstream, normal, and valued society members. From the point of view of a labelling theory of stigma, the construction of the label in itself creates and defines deviance (Becker 1963; Scott and Douglas 1972). Consequently, the labels of 'homeless' and 'HIV/AIDS', attached to identified individuals and groups, are evaluative rather than merely descriptive or neutral classifications, and indicate to the wider public the undesirability of such individuals and groups.

The meaning of the labels 'homeless' and 'HIV/AIDS' takes shape for the public through various avenues: their own personal experience, widespread media exposure, and the information disseminated by experts and policy-makers (Lee *et al.* 1990). Moreover, Gans (1995) argues that in addition to these avenues labelling is developed using what he calls 'imagined knowledge', or those sets of judgements based on stories which coincide with the prejudices of those who judge (p. 11). Since the stigmatization of specific groups is often more about perceived threat than material experience, the production of the labels 'homeless' and 'HIV/

AIDS' becomes central in the public understanding of homelessness and HIV/AIDS.

Stigmatized labels are for the most part categories imposed from outside the stigmatized groups. That is, in the stigmatization processes associated with homelessness and HIV/AIDS, homeless persons and people living with HIV/AIDS have little to do with the construction of stigmatized labels. Labels are expressions of society's contemporary social and moral boundaries, and are useful in controlling individuals and groups not fulfilling social norms. That is, once individuals are labelled 'homeless' or 'HIV-positive' and enter the social service network to access material and financial support, they become prone to organizational rules and procedures. The dependence on social and human services by homeless persons and people living with HIV/AIDS means an inevitable invasiveness and control by organizations, bureaucracies, and individuals providing such services (Young 1990). The labelling of individuals as 'homeless' or 'living with HIV/AIDS' is a necessary process for those individuals to access programmes (because one cannot access homeless services if one is not 'homeless') and in so doing creates the conditions for social control through bureaucratic institutions.

Not only do labels perform normalization functions for society's members, but the development of labels fulfils specific professional needs as well. As Young (1990) argues, 'welfare agencies construct the needs themselves' (p. 54). That is, in constructing the labels 'homeless' and 'HIV/AIDS', the medical and helping professions develop categories and identifications requiring professional intervention, and indicate how such intervention should take place. Mental health care has provided clear examples over time about the role of professional organizations and the changing nature of labelling, such as the American Psychiatric Association and their literal classification of mental disorders in the *Diagnostic and Statistical Manual of Mental Disorders*.

The practice of defining and naming mental disorders by professional organizations illustrates the bureaucratization of stigma. Bureaucratization takes shape through institutional and professional practice, rules, and laws. In particular, in their defining the nature of homelessness and HIV/AIDS and their responses to these two conditions, researchers, advocates, human service providers, and policy-makers develop a system of 'codification, definition of offences, the fixing of a scale of penalties, rules of procedure, [and] definition of the role of magistrates' (Foucault 1977: 102; also de Certeau 1986; Schur 1971). Thus, not only are experts and policy-makers working to delineate the characteristics and causes of homelessness and HIV/AIDS to develop appropriate governmental responses, but such actions also create partitioning within society, a ranking of its members, and the authority of such experts to devise and enforce social rules (Giddens 1984). In terms of HIV/AIDS, for example, the identification of individuals shifts from 'HIV-positive' to 'AIDS' with the changing count of specific cells, which indicate the body's level of immunity to disease and opportunistic infection. Such identification as 'HIV-positive' or living with 'AIDS' (which has changed with growing scientific knowledge concerning the nature and progression of HIV/AIDS) not only has implications for how individuals are dealt with by the wider public, but is also central in accessing medical treatments and monetary resources provided by the government. Government control over the definition of what constitutes 'HIV' and 'AIDS' has been particularly salient in the past non-recognition of gynaecological conditions associated with HIV/AIDS, and the consequent lack of study and programme development centred on women living with HIV/AIDS (Patton 1994).

The labels 'homeless' and 'HIV/AIDS' carry differing degrees of stigma along a set of dimensions, resulting in varying acceptability shown towards particular homeless persons and people living with HIV/AIDS. Once an individual is labelled as 'homeless' or 'living with HIV/AIDS', the stigma attached to that person is more/less extreme given perceived and actual degrees of: functionality (e.g. more functional individuals are deemed more acceptable than less functional persons); aesthetics (e.g. the degree of visibility, effort to conceal devalued characteristics, and social acceptability of appearance); interference with established rules for social interaction (e.g. more interference means lesser acceptance); attributable blame or personal responsibility for the stigmatized condition; unpredictability and danger posed; how 'curable' or immutable is the stigmatized condition; empathetic nature of the condition; and contagiousness (Dear *et al.* 1997; also Jones *et al.* 1984; Scott and Douglas 1972).

Label development regarding homelessness and HIV/AIDS is time and place specific. Over the past decade, the clarification of the labels has centred on defining the concentrations of people in particular places, patterns of migration, their service needs, and the causes of homelessness and HIV/AIDS (Takahashi 1996). The identification of causes of homelessness and HIV/AIDS is particularly important because discipline or social punishment for these two conditions tends to be adjustable, depending on whether homelessness and/or HIV/AIDS are due to circumstances or individual intent (Foucault 1977).

The label of 'homeless' has increasingly become synonymous with various forms of deviance over time, including laziness, alcoholism, drug abuse, mental disability, criminalism, and even perversion (Gallagher 1994). But these meanings of the label 'homeless' are not monolithic; they depend in large degree on where people attribute the causes of homelessness, both in general and for specific individuals. When homelessness is viewed as being caused by structural change (especially the fallout caused by economic restructuring, such as loss of employment through lay-offs and plant closures) or natural disasters (such as wildfires, hurricanes, tornadoes, or earthquakes), members of the public often perceive the homeless as victims of circumstance and consequently advocate government intervention in providing housing and employment. However, if homelessness is viewed as emanating from individual actions (such as substance abuse or the lack of stable employment), then more coercive and disciplining actions are often called for (e.g. the denial of welfare benefits).

There are similar implications in terms of the public's response to the label of 'HIV/AIDS'. On the one hand, when HIV/AIDS is understood as due to abnormal or deviant behaviour (such as homosexual activity or intravenous/injecting drug use), then the label 'HIV/AIDS' is likely to elicit a high degree of devaluation and discipline. On the other hand, when HIV/AIDS is seen as illness there is a greater likelihood that the public response will parallel responses to other terminal conditions. In the recent past, both responses have resulted in public rejection of persons living with HIV/AIDS, but with greater scientific information being disseminated throughout the populace, there has emerged greater differentiation by the public in their perception of persons living with HIV/AIDS depending on the mode of transmission. Even with such differentiation, the label of 'HIV/AIDS' has tended to focus public and scholarly attention on the virus rather than on individuals. For instance, in monitoring the concentrations of persons living with HIV/AIDS, their mobility patterns, and the sources of transmission, research has largely centred on the spatial concentration and mobility of the virus. Thus, through the construction of the label of 'HIV/AIDS', people (and in particular their bodies) lose their individuality, becoming dehumanized as they are identified as the vector of HIV/AIDS (Brown 1995; Epstein 1995).

But emphasizing solely the labelling mechanism in producing stigma overlooks the complexity underlying the social relations underlying stigmatization. As Foucault (1977) has argued, 'The art of punishing . . . [rests] on a whole technology of representation' (p. 104; also Treichler 1988). The boundaries defined by the stigmatization of homelessness and HIV/AIDS are not simply negative labels attached to individuals and groups by powerful individuals and organizations. Nor is stigmatization monolithic and unchanging. Instead, the nature of stigma is fluid and continually contested, with the stigmatization of homelessness and HIV/AIDS constantly being constructed, transformed, and managed. The fluidity and change in stigma may work to reinforce current hierarchical relations of power (in terms of continuing to marginalize specific groups) or may work to challenge existing labels and demarcations. Schur (1980) has referred to this fluidity in the social relations of stigma as 'stigma contests' or 'the continuing struggles for competing social definitions' (p. 3).

We can conceptualize this fluidity in the stigmatization of homelessness and HIV/AIDS in terms of three interrelated dimensions: non-productivity, dangerousness, and personal culpability. These three dimensions constitute perceived and material threats to the social relations of production and reproduction (Takahashi 1996; also Spitzer 1975). The first dimension, the perceived and material degree of non-productivity, has been central in contemporary definitions of stigma, particularly with the recent rise in homelessness and HIV/AIDS. The privileging of productivity is a ubiquitous social norm. As Piven and Cloward (1977) have argued, 'Ordinary life for most people is regulated by the rules of work and the rewards of work which pattern each day and week and season' (p. 11). If society defines acceptability by whether individuals work in the formal and paid labour market, then people who are perceived as not working (e.g. homeless persons) or who might be blocked from the labour market (e.g. persons who identify themselves as HIV-positive) become relegated to less valued positions in society.

The reason for the ubiquitous nature of (non)productivity as a social norm stems from the existing structure of capitalist social relations. These relations, constituting the particular economic and political structure of capitalist society, develop from a combination of the activities by economic subjects, non-economic conditions and individuals, and historical production processes (Clark and Dear 1984). To maintain and reproduce this system, productive individuals who participate in the paid labour market or engage in activities supportive of such systems (such as unpaid domestic labour) are accorded relatively privileged status in comparison to seemingly non-productive persons such as homeless individuals or persons living with HIV/AIDS. Non-productive persons, which 'the system of labour cannot or will not use' (Young 1990: 53) because they are unacknowledged as being valuable by wider society, are viewed as threats to collective consumption and communal life and as drains on productive members of society (Cox 1989; Jones *et al.* 1984; Piven and Cloward 1971). Particularly with respect to homelessness, these threats to collective consumption and communal life become expressed as attributions of laziness, undeserved leisure, and dependence on society (e.g. 'cheating').[8]

The significance of productivity in the definition of acceptability and stigma has become critical with the emergence of homelessness and HIV/AIDS. Research over the past decade, both in homelessness and HIV/AIDS, has suggested that the growth in these two populations is largely concentrated among individuals in their twenties and thirties (Selik *et al.* 1993; Wolch and Dear 1993). Thus, populations who are homeless and/or living with HIV/AIDS are often in their prime productive years, and, therefore, homelessness and HIV/AIDS constitute conditions or attributes which prevent participation in the paid labour force, and often even in tasks which might support production processes (such as unpaid domestic labour).

The second dimension of the social relations of stigma, danger and unpredictability, has long been associated with the devaluation of specific groups (Douglas 1966; Gilman 1988). Indeed, the characterization of homeless persons and people living with HIV/AIDS as dangerous and unpredictable has been critical in their stigmatization. Robert A. Scott (1972) has argued that dangerousness and unpredictability play such a central role 'because they are potentially more powerful than the order that stands against them' (p. 21). With respect to homelessness, for example, there is the widespread perception that many of these individuals are mentally unstable and criminal (Gallagher 1994; Hopper and Baumohl 1994). The perception of mental instability is due in large part to the widespread understanding of homelessness as being the result of deinstitutionalization policies enacted during the 1950s and 1960s in the US, the widely published statistic that about one-third of homeless individuals are mentally disabled, and the highly visible and seemingly abnormal behaviour of specific homeless persons (Blasi 1994; Fink and Tasman 1992; Wolch and Dear 1993).

Criminality is also commonly attributed to homelessness. A lack of participation in the paid labour force and lack of available and accessible public assistance means that many homeless persons must participate in the informal and illegal economies for daily survival (Takahashi and Wolch 1994). Although it is true that many homeless persons have been jailed, with a proportion for violent crimes, much of this incarceration is traceable to trespassing and public nuisance. In addition, the social construction of homeless as criminal has increasingly been institutionalized through local ordinances banning activities engaged in by homeless persons (Takahashi 1996). Anti-homeless laws in general (such as anti-camping, anti-aggressive panhandling, and other legislation) have contributed to the impression that homeless persons are involved in illegal activities which are dangerous (or at least potentially dangerous) to life and property.

Danger, whether actual or potential, has also largely defined societal understanding of persons living with HIV/AIDS. Such danger is very much linked to the fear of contagion. Although there is scientific evidence that HIV/AIDS cannot be transmitted through casual contact (e.g. shaking hands, kissing, and hugging), there is still the potent fear that HIV/ AIDS constitutes a threat if persons living with HIV/AIDS are in close proximity. Danger associated with HIV/AIDS also extends beyond the potential contagion associated with the risk of transmission (Sontag 1988). There are also the dangers posed to established social norms by individuals living with HIV/AIDS. There is still the widespread belief that HIV/AIDS remains largely confined to the homosexual male population and intravenous/injecting substance users. With such beliefs, public perception about danger associated with HIV/AIDS also stems from the influence such individuals might have on established social norms (Patton 1990, 1994). For example, communities might view persons living with HIV/ AIDS as not only representative of a terminal, incurable, and still mysterious condition, they might also see such individuals as threats to the established religious, cultural, and other norm structures which serve to consolidate households and families. Thus, not only might people living with HIV/AIDS be perceived as threats to people coming in direct contact, they might also be seen as moral dangers and threats to the overall social fabric of local communities (Epstein 1995; Patton 1990; Sibley 1995; also Douglas 1966).

The third dimension of stigma which is particularly significant for homelessness and HIV/AIDS is personal culpability, or how much individuals are seen to be responsible (and thus to blame) for their circumstances. Personal culpability comes into play because a lesser amount of blame attributable to individuals and groups may offset the lack of physical or behavioural aesthetic. That is, conditions such as homelessness or HIV/AIDS may be more forgivable if they are seen to be not caused by individual choice. Instead, if homelessness and HIV/AIDS are seen as an outcome of deliberate or intended individual behaviour, such individuals are likely subjects for societal discipline or punishment. As Young (1990) argues, 'To blame an agent means to make that agent liable to punishment' (p. 151). Personal culpability has been linked to the differentiations made between many 'deserving' and 'undeserving' groups. There is, for example, an extensive literature on this issue and the culture of poverty and the undeserving poor (Feagin 1975; Harrington 1981; Katz 1989).

The degree of personal culpability for becoming homeless has remained a vital element in societal understanding of the causes and possible solutions for homelessness (Baumohl 1989; Dear and Gleeson 1991). When the definition of homelessness is primarily confined to individual deficiency, the public is much more likely to blame homeless individuals for their becoming and remaining homeless. Conversely, if homelessness is understood as emanating from other than individual sources, the public might be less likely to blame individuals for their homelessness. While recent attitudinal studies in the US have indicated that there is a wider acknowledgement among the populace that homelessness derives in large part from structural and institutional changes which homeless individuals could not control (Gallup 1995; Lee *et al.* 1992; Toro and McDonell 1992), there remains the perception that many homeless persons are to some degree personally responsible for their becoming homeless, or, more importantly, for their remaining homeless (e.g. Baum and Burnes 1993). The greater the perception that individuals are personally culpable for becoming or remaining homeless, the less acceptable they tend to be to community residents.

Views about personal culpability also influence people's evaluations of the acceptability of persons living with HIV/AIDS. Public perceptions concerning HIV/AIDS are still largely influenced by the metaphor of deviance, implying wilful behaviour on the part of the person living with

HIV/AIDS. Persons living with HIV/AIDS are often viewed as being personally responsible for acquiring the condition (Blendon and Donelan 1989). The perception of personal responsibility for acquiring HIV/AIDS is strongly related to the well-publicized epidemiology of HIV/AIDS, especially the relatively high risk of transmission among men who have sex with men and intravenous/injecting substance users (Epstein 1995; Patton 1990). There is a contradictory dimension inherent in public perception concerning the personal culpability involved in the transmission of HIV/AIDS. On the one hand, as already mentioned, as a reflection of the dangerous dimension of stigmatization, people fear that close proximity to persons living with HIV/AIDS (i.e. proximity to the embodiment of the condition) will enhance their probability of infection or the possibility of moral degradation. On the other hand, people tend not to attribute proximity of HIV/AIDS as the explanation that persons have become HIV-positive or are living with HIV/AIDS. Rather, these individuals are often seen as having participated in deviance (whether homosexual activity, promiscuity, or intravenous/injecting substance use), and are therefore seen as personally culpable for having been infected.

The fluidity of stigma can be captured by understanding these three dimensions as continua rather than categories. In this way, the public's perception of the degree of non-productivity, the degree of dangerousness, and the amount of personal culpability associated with homelessness and HIV/AIDS can be mapped on to a three-dimensional continuum (Fig. 3.1). The intersection of these three continua define cognitive 'spaces of stigma' which reflect the variation in acceptance and rejection exhibited by the US public towards homelessness and HIV/AIDS. The least stigmatized cognitive spaces (and therefore the most acceptable) lie closest to the intersection which reflects being productive, non-dangerous, and blameless. Individuals who can be characterized in this space of stigma are seen as being contributing members to the local community and to society at large. They are the productive members of society, working for wages, paying their taxes, volunteering for neighbourhood activities and other social causes, and not exhibiting significant and visible evidence of deviance from social norms. Groups existing in the space defined by non-productivity, non-dangerousness, and being blameless may also be evaluated positively because they are not significantly dangerous. Young children, for example, occupy this cognitive space of stigma, since they are relatively unproductive, but non-dangerous, and not considered personally culpable for their existence.

In this three-dimensional continuum of stigma, homeless persons and people living with HIV/AIDS occupy a cognitive space defined by non-productivity, dangerousness, and personal culpability. Since not all of these persons are deemed entirely non-productive, always dangerous, and ultimately personally culpable for all of their problems, the population of

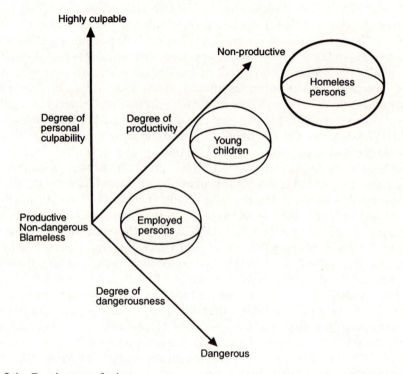

Fɪɢ 3.1. Continuum of stigma

homeless persons and people living with HIV/AIDS occupies a *space* of
stigma, rather than a point of stigma defined only by the extremes of these
negative evaluations. In other words, the boundary of the cognitive space
of stigma occupied by homeless persons and people living with HIV/AIDS
may span a significant range along the unproductive–productive,
dangerous–nondangerous, and personally culpable–blameless dimensions.
The space of stigma occupied by homeless persons and people living with
HIV/AIDS translates into the variations in accepting and rejecting
responses by the public in different spatial, socio-demographic, political,
and historical contexts.

Conclusion

The critiques brought forward by ongoing social theory debates concerning
difference beg a more nuanced and contextualized conceptualization of the
NIMBY syndrome towards human services than has previously existed. I
have argued in this chapter for a conceptual formulation of the NIMBY

syndrome based in theoretical understandings of difference and the social relations of stigma. As a growing literature on postmodern, post-colonial, poststructuralist, and other post-concepts and critics of such frameworks have pointed out, however, rejecting universalism and dualisms does not necessarily translate into greater inclusion (nor for that matter does essentialism always result in exclusion). And in the case of the NIMBY syndrome, more and diverse communities are identifying themselves as marginalized and excluded (even suburban home-owning residents), so that even the notions of difference and exclusion are becoming relative terms strategically used to further specific agendas.

But difference does offer opportunities to understand further the widespread rejection across diverse communities towards homelessness and HIV/AIDS, beyond the common representation of such beliefs and practices as purely the selfish, exclusionary reactions of residents and municipalities. To develop further the theoretical perspective presented here, the concept of the socio-spatial stigmatization of homelessness and HIV/AIDS will be explored in greater depth in the following two chapters. The first of these chapters will explore how stigma becomes attached to particular persons and places. While much of the scholarship on stigmatization has emphasized the devaluation of persons, especially with respect to homelessness and HIV/AIDS, there has been much less emphasis on the linkages between the stigmatization of persons and places. The role of human service provision in the devaluation of persons and places is central to Chapter 4.

Following this discussion is the related issue of the role of race and gender in the stigmatization of homelessness and HIV/AIDS. This issue is taken up in the final chapter in this section on conceptual understanding of the NIMBY syndrome. Although geographers and planners have increasingly confronted the problematic character of both race and gender, such discussions have not generally permeated the scholarship focusing on homelessness and HIV/AIDS or on public response to these two social crises. Instead, discussions of race and gender within substantive explorations of homelessness and HIV/AIDS tend to focus on differentiating homeless persons and people living with HIV/AIDS by racial or ethnic categories. Such differentiations are often made with the intent of identifying at risk populations or some distinct differences in the type of homelessness or HIV/AIDS experienced by these identified groups. To move the debate beyond these characterizations by race or gender, Chapter 5 discusses how the problematic of race and gender constitute significant facets of the fluidity of stigma concerning homelessness and HIV/AIDS.

4

Assignation of Stigma to Persons and Places

HOMELESSNESS and HIV/AIDS have encountered increasing stigmatization over the past decade in the United States. As the previous chapter indicated, the social relations of stigmatization make the NIMBY syndrome much more complex than solely reactionary, selfish responses by intolerant individuals. Rather, the NIMBY syndrome is evidence of the social and moral boundaries emanating from the stigmatization of homelessness and HIV/AIDS. As such, community opposition towards human service facilities is part of a broader set of practices delineated by labelling, avoidance, discrimination, and even violence. But community response is not monolithic; it is framed by the fluid nature of stigma (which I described in Chapter 3 using three dimensions: non-productivity, dangerousness, and personal culpability). The question thus arises, how do such social definitions of difference adhere to particular individuals and places?

This chapter considers the means by which specific persons and places become and remain stigmatized. In terms of the stigmatization of persons, the chapter outlines three mechanisms which are particularly important with respect to homelessness and HIV/AIDS: the use of stereotypes and depersonalization, the identification as belonging to risk groups, and the development of the personal identities of homeless persons and people living with HIV/AIDS. With respect to the devaluation of places, the chapter explores the person–place dynamic underlying the stigmatization of homelessness and HIV/AIDS. Specifically, I explore the mechanisms underlying how places inherit the stigmatization of persons, how persons may inherit the stigmatization associated with places, and how such stigmatization (both of persons and places) may be circumvented or challenged.

Whether stigma is conceptualized as normalization, moral boundaries, social control, or difference, the assignation of such stigma to particular individuals has been a fundamental emphasis of social science research on this issue. However, less attention has been paid to the ways in which place influences and interacts with stigmatization. The issue of the person–place dynamic in the stigmatization of homelessness and HIV/AIDS is extremely important for understanding the NIMBY syndrome with respect to human service facilities. The particular location and the sphere of influence of proposed human service facilities have often been the igniting catalysts sparking localized opposition (Dear and Taylor 1982). However, the person–place dynamic inherent in stigmatization is more complex than spatial location; it involves the incorporation of the nuances of place with the social relations of stigma. Understanding the contemporary and future expressions of community response to homelessness and HIV/AIDS requires an appreciation of such linkages, affected by local experience and policy responses to homelessness and HIV/AIDS.

How Stigma Adheres to Particular Persons

It is common to think about the devaluation of persons when discussing stigmatization. The literature on the stigmatization of persons and groups has a long history and has engaged many distinct conceptual frameworks. In explaining stigma, social science researchers have often focused on stigmatized groups, such as sex workers, white-collar criminals, and mentally disabled persons (e.g. Ellis 1987; Traub and Little 1985). Such studies, many conducted under the substantive rubric of 'social deviance' research, have highlighted various mechanisms by which stigma adheres to particular individuals and groups. Of central importance in this body of research is the development of stereotypes and the process of depersonalization inherent in stigmatization.

Stereotypes

The construction of stereotypes constitutes a mechanism to legitimate and reinforce use of the labels 'homeless' and 'HIV/AIDS'. Stereotypes differ from labelling or categorization in the following sense. Labelling or categorization refers to the 'process of lumping together different social and physical objects into a single category' while stereotypes imply 'a presumed correlation between one trait (category membership) and other traits (e.g. personality and physical characteristics' (Jones *et al.* 1984: 156–7). In stereotyping homeless persons and people living with HIV/AIDS, the judge generates a set of personality, physical, and behavioural characteristics about individuals from the labels 'homeless' and 'HIV/AIDS'. In so doing, distinctions can be made between 'the "good" and "bad" world and self, between control and loss of control, between acquiescence and denial' (Gilman 1985: 17). This process of generating characteristics results in the negative assessments of specific groups. Gans (1995) characterizes negative stereotypes in this way: 'they may extrapolate from small kernels of truth about some people to large imagined untruths that are applied to everyone in a group' (p. 12).

A central key to successful stigmatization is the ability of stereotypes to obfuscate the deliberate social construction of such labels, to make the negative connotation of 'homeless' and 'HIV/AIDS' the responsibility of the individual (Ainlay and Crosby 1986). The resulting reification of the labels 'homeless' and 'HIV/AIDS' means that the public begins to believe that there exist homogeneous groups of homeless people and persons living with HIV/AIDS who reflect the current definitions of homelessness and HIV/AIDS. For those in decision-making positions, such stereotypes of homelessness and HIV/AIDS are particularly important given that few advocates, service providers, planners and policy-makers have directly experienced this 'social space' (Beauregard 1995: 67). In a paradox of visibility, being marked through stereotypes makes visible the devalued character of a person and their group while at the same time rendering them invisible as subjects (Blunt and Rose 1994; Epstein 1995; Golden 1992; Young 1990). Homeless persons on the streets of large cities embody this paradox. Passers-by immediately recognize the proximity of homeless persons, because of physical and behavioural markers (e.g. clothing, possessions, soliciting monetary contributions), but often, these passers-by do not recognize homeless persons' existence, often looking away or ignoring their queries for money, and reacting as if these homeless persons were invisible or non-existent.

The devaluation associated with stereotyping exacerbates often already marginal material circumstances, reinforcing the devalued status and role of stigmatized individuals and further depressing future life chances. The devaluation of status interacts with the concomitant depression of life chances resulting in the additional evaluation of stigmatized persons as undeserving of help. As Gans (1995) argues, 'the reification of a label . . . usually leads to the assignment of *moral* causality' (p. 62, emphasis in text). This characterization as undeserving results from the combination of devalued social status and the limited ability to improve life circumstances, resulting in accusations of laziness and societal dependence. The reified stigma of homelessness and HIV/AIDS thus devalues social status and minimizes people's ability to improve their quality of life.

In stereotyping homelessness and HIV/AIDS, individuals become depersonalized in the view of wider society. For example, homeless persons take on the deviant characteristics of homelessness generally—mental disability, substance abuse, and criminality (Rocha and Dear 1989). Depersonalization 'implies that the individual is nothing but an instance of the discredited category, and, correspondingly, that all members of the category are basically alike' (Schur 1980: 147). In becoming a member of a discredited category, whose members are in essence alike, homeless persons and people living with HIV/AIDS are deemed essentially different from everyone else, effectively marking them as Other. As Young (1990) argues, '[g]iven the normality of its own cultural expressions and identity, the dominant group constructs the differences which some groups exhibit as lack and negation' (p. 59).

The stereotypes of the Other have historically been evaluated across multiple dimensions. Young (1990) categorizes these as physical health, moral soundness, and mental balance. Those with lesser or non-aesthetic physical health conditions, less than acceptable moral behaviour, and mental instability are deemed Other and experience stereotyping associated with deviance. In assigning negative evaluations to those considered Other, the body of deviants become widely associated with dirt, ugliness, craziness, disease, and homosexuality (Douglas 1966; Foucault 1973; Gilman 1985). We constantly conduct such assessments of people around us, judging which persons and behaviours are acceptable or not, and adjusting our behaviour accordingly; 'Everyday life is, among other things, a never-ending flow of moral surveillance' (Gans 1995: 11).

What are the Signs of the Other and Difference?

Indicators of difference and deviance are often assessed from a distance, through highly visible appearance and behavioural characteristics deemed to emanate from homelessness and HIV/AIDS. In terms of homelessness, common means of identification include appearance and aesthetic (such as clothing, cleanliness, hygiene, bodily disfigurement, and physical health), and visible behaviour (such as perceived loitering, wandering, panhandling, pushing a shopping cart filled with personal items, and talking to oneself).[1] One need not exhibit all of these characteristics to be assessed homeless, as long as the homeless person is expected to be seen at a particular time and place, and that homeless person fulfils enough of the stereotypical characteristics for the judging individual to be satisfied with her or his assessment.

Even though a person characterized as homeless or living with HIV/AIDS has many more identities than this sole characterization, the identification as stigmatized has significant inertia and tends to be more important than any other identities (Berger and Luckmann 1966). All people have affiliations to multiple groups, not only stigmatized ones, yet after becoming labelled 'homeless' or 'living with HIV/AIDS', the stigmatized identity takes precedence. Behaviours and appearances which might be deemed eccentric or strange by otherwise non-stigmatized, normal persons take on special deviant meanings when people are stigmatized; 'because the supposedly deviant status takes on a special saliency, people (and perceived problems) are "reread" in terms of ancillary qualities believed to be characteristics of a given type of presumed deviator' (Schur 1980: 146).

The defining characteristics associated with the stereotype of 'homeless' and 'HIV/AIDS' are constructed in part through the rules and practices of professions and institutions associated with homelessness and HIV/AIDS (Snow and Anderson 1993). There are several facets of rules and practices which create, manipulate, and reinforce the stigma assigned to specific individuals. One pragmatic facet of such practices lies in organizational maintenance and growth, bureaucratic legitimation, and the form and function of the state. The reproduction of categories such as 'homeless' and 'HIV/AIDS' are necessary for the maintenance of professional status, funding, and providing services. Without the continued existence and definition of the label 'homeless' and the needs that this label implies for professional intervention, for example, the growth industry characterizing human service delivery associated with homelessness would slow considerably (Wolch and Dear 1993). This growth industry emerges from the new economies and employment opportunities offered by collective stigma (Schur 1980). Likewise, the label 'HIV/AIDS' has associated with it millions of dollars in private, non-profit, and public resources invested in medical research and data gathering, education and prevention programmes, and human service delivery. In part, the growth in these newly emerging human service economies stems from the conceptualization of the bodies of homeless persons and people living with HIV/AIDS as vectors for homelessness and HIV/AIDS (Brown 1995; Foucault 1973). As such, characterizing and counting the bodies of homeless persons and people living with HIV/AIDS constitute primary functions of public policy, research, and human service delivery interested in showing the increasing significance of these two conditions. The regulation of the definition of 'homelessness' and 'HIV/AIDS' therefore become an integral part of professional practices and bureaucratization, indicating how the representation of issues is fundamentally related to the policies formulated to resolve them (Berger and Luckmann 1966; see also Cockburn 1977; Hoch and Slayton 1989; Liggett and Perry 1995; Piven and Cloward 1971; Wagner 1993).

Stereotypes, the State, and Bureaucratization

The construction of 'homeless' and 'HIV/AIDS' also generally supports the form and function of the state. Through the ongoing characterization of 'homelessness' and 'HIV/AIDS' as stigmatized (e.g. non-productive, dangerous, and personally culpable), the state 'secure[s] the conditions for capital accumulation and the reproduction of the social formation' (Clark and Dear 1984: 37; also Cockburn 1977). Although there is ample inter-level conflict concerning the implementation of welfare reform, for example, the state's role in the enforcement of stigmatization is clear in its management of welfare policy.[2] Contemporary efforts at welfare reform in the United States, for example, have made a lack of participation in the paid labour force punishable by a withdrawal of state support (via income maintenance through programmes such as Aid to Families with Dependent Children, or AFDC, and medical insurance such as Medicaid). Under federal guidelines that took effect on 1 October 1996, renaming the AFDC programme Temporary Assistance to Needy Families, 'most people on welfare can receive benefits for a maximum of five years during their lifetime, after which neither the state or the federal government is obligated to offer assistance no matter what their circumstances' (Rivera 1996: A24). Growing public outcry about the undeserving nature of welfare recipients has encouraged public officials to continue the process of welfare reorganization which began in the 1980s. Local states are also acting autonomously to discipline homeless persons by criminalizing outdoor camping and income generating activities such as panhandling. Such practices by local states have been prompted in part by businesses arguing that the presence of homeless persons is bad for business.

However, such state practices are not unproblematic in their design and implementation. Intergovernmental conflict resulting from the delineation of responsibility for the indigent has caused cities, counties, states, and federal agencies to dispute the appropriate level of fiscal and programmatic responsibility. In California, for example, the state shares responsibility for providing income maintenance and in-kind services with the federal government (recently through Aid to Families with Dependent Children and Medicaid, and, with the passage of welfare reform federal legislation, through Temporary Assistance to Needy Families). Municipalities in California are responsible for providing income support and services for the indigent in their jurisdictions, through welfare programmes known as General Assistance or General Relief. However, there has been ongoing conflict about the appropriate level of responsibility; such conflict is often expressed and implied through the changing levels of financial support provided through the various welfare programmes. The state of California, for example, to address the unequal costs of living experienced in urban and rural jurisdictions, has developed a geographically based allocation system, where persons in cities are allocated slightly more money monthly than those living in rural areas. Such policy practices interact with service provision creating places with service rich environments and those with service poor environments (Dear *et al.* 1994).

A second facet of professional rules and practices is that stigmatized persons affected by stereotypes and labels have little to do with their construction and maintenance. The construction of the definitions of homelessness and HIV/AIDS, for example, emanate from the professional worlds of research, public policy, and service provision for the most part, without consultation or inclusion of the views of homeless persons or people living with HIV/AIDS (Wagner 1993). In addition, as part of the accessing and use of the social services system, stigmatized persons become subject to rules and procedures over which they play a minimal role. Thus, stigmatized persons not following rules and procedures outlined by professions and institutions reinforce their identification as deviant and abnormal and may even be denied access to the system of human services. To reap the benefits offered by the system of human services, persons must become identified as 'homeless' or 'living with HIV/AIDS', enter and negotiate the web of human service delivery, and follow the rules and procedures outlined by providers (Dear *et al.* 1994). The bureaucratization of human service delivery has made rules and procedures very important to the stigmatization associated with homelessness and HIV/AIDS. The importance of following the rules, and subsequent discipline or punishment for not following procedures, is due to what Young (1990) calls 'the universalization and standardization of social or co-operative activity' (p. 77). Discipline and stereotyping interact as the negative characteristics associated with the stereotype of homeless or living with HIV/AIDS are often enabled, created, or perpetuated by service providers through the interpretation of past and present behaviour as manifesting deviance (Goffman 1963; Schur 1980).

A third and extremely important facet of professional practices in relation to stereotyping is the widening of the sphere of intervention from groups actually homeless and living with HIV/AIDS to wider groups at risk.[3] Not only do professional practices manage the stereotyping of those deemed homeless or living with HIV/AIDS, but with the identification of risk groups this professional influence extends to those who might become homeless or HIV-positive. The development of individuals and groups at risk of homelessness and HIV/AIDS has become an important method for advocates, human service providers, and policy-makers in preventing homelessness and HIV/AIDS. On the one hand, in identifying and targeting groups which are not already homeless or living with HIV/AIDS, advocates, human service providers, and policy-makers are working to identify risk factors, risky behaviours, and environmental/structural factors which can be manipulated to prevent a rise in the incidence of homelessness and HIV/AIDS. As such, the characterization of at risk groups is deemed a necessary step for resolving the social crises of homelessness and HIV/AIDS. It has comprised the primary tool for prioritizing funding and programmes for persons and groups at risk of homelessness and HIV/AIDS (Goldin 1994). On the other hand, the use of risk groups creates a 'tendency to distance the "general population" from "risk groups" . . . facilitating public definitions of the HIV epidemic as a problem which concerns others, not oneself and one's own "group"' (Schiller *et al.* 1994: 1344).

In addition, in categorizing risk groups, people become depersonalized to the extent that '[b]odies only differ according to their risk capacity for [HIV] transmission' (Brown 1995: 164). In identifying what constitutes risk of becoming homeless or HIV-positive, researchers and professional human service providers have often looked to cultural practices which encourage risky behaviour. Thus, in terms of HIV/AIDS, the deviance and promiscuity associated with gay culture serves to cement the notion that homosexuality constitutes a subculture with high risk of HIV transmission. The widely accepted association of male homosexual/bisexual contact with the identification of high risk of HIV transmission minimizes the wide variation in the sexual practices of gay men, the high risk of specific heterosexual practices, and the lack of recognition of male to male sexual contact as homosexual by many groups (Carrier and Magaña 1992). In addition, the categorization of risk groups with respect to HIV/AIDS tends to be racialized and class specific. As Schiller *et al.* (1994) argue, 'the literature on HIV infected intravenous drug users tended to be of "minorities," or "blacks and Hispanics," while homosexuals were seen as white and middle-class' (p. 1340). Thus, not only does the declaration of specific groups as at risk serve to delineate difference between those groups and the wider population, but the reification of risk groups means that the wider population becomes much less concerned about homelessness and HIV/AIDS (i.e. because homelessness and HIV/AIDS will only impact those already homeless or living with HIV/AIDS or those at risk).

Self-Identity Formation

While the labelling and stereotyping of individuals by professionals and institutions are central in the stigmatization of homelessness and HIV/AIDS, the construction of self-identity as homeless or living with HIV/AIDS is also vital in the dynamic underlying the social relations of stigma. Individuals with stigmatized characterizations directly inherit the stigma defined through labels, stereotypes, and risk groups, but, in addition, may also internalize these marginalized or devalued identities. Not only then does society define stigmatized individuals primarily through their stigmatized identity, but those individuals may do so as well, minimizing their identification with other groups. As Anderson (1991) has argued, 'those on the receiving end of identity classifications come to live within the paradigm fashioned by their oppressors and define their own identities accordingly' (p. 27; also Edgerton 1967). Those identified as homeless or living with HIV/AIDS, because of the cognitive influence of stereotyping and labelling and the material need for human services, begin to adapt their life paths and daily routines to accommodate these stigmatized identities. Thus, labels and stereotypes 'may sometimes force the labeled to behave in ways defined by and in the labels' (Gans 1995: 12). Those stigmatized persons who can successfully accommodate this identity, and access human services and other resources, may solidify this new self-identity though daily routines and practices. For homeless persons, for example, developing the ability to cope with daily needs may tend to reinforce individual identity as 'self as homeless' or 'self as recipient' (Rowe and Wolch 1990: 199).

In identifying individuals as homeless or living with HIV/AIDS such individuals may gain access to human services dedicated to these populations, but in becoming part of the reified identity of homeless or living with HIV/AIDS, they also undergo devaluation in social status and material circumstance. The changes in daily routines produced by identification as homeless or living with HIV/AIDS act to reinforce the perceived and material stigmatization experienced by individuals. This is not to say that stigmatization alone causes the reduction in life chances or material circumstances. Homelessness and HIV/AIDS bring with them clear material changes in life circumstances (e.g. economic marginality, depressed physical and mental health, and limited access to material and emotional support from family and friends). But the identification of individuals as homeless or living with HIV/AIDS interacts with this marginalization, limiting the ability of individuals to become or remain part of wider society. After becoming homeless or after diagnosis of HIV/AIDS, life circumstances and self-identities often destabilize as societal devaluation takes place. For homeless persons, for example, the frequent substitution of social support network members 'demands immediate attention and diverts energies away from long-term strategies for re-entry into the homed mainstream. The result is prolonged homelessness and a transformation in identity and self esteem' (Rowe and Wolch 1990: 199).

The importance of daily paths and routines implies the centrality of place in the production and reproduction of social relations of stigma. Stigma is embodied in homeless persons and people living with HIV/AIDS, and is emplaced through the built environment and broader spatial relations. But place is not merely an inheritor of the stigmatization ascribed to persons. There is a dynamic and dialectic relationship between stigmatized identities and stigmatized places, where stigma is transferred to places from persons and vice versa, and where the construction of stigma must be seen as Edward Soja (1989) argues, 'as simultaneously contingent and conditioning, as both outcome and medium' (p. 58).

Stigmatizing Places

The social construction of stigma concerning homelessness and HIV/AIDS is informative about how the public might respond towards homeless persons and people living with HIV/AIDS. However, in any effort to understand community response to human services, there is the added complication of the built environment and the larger spatial relations defined by human service users and the communities in which they are located (Laws 1994; Ruddick 1996; Takahashi 1997). Relph (1976) describes this socio-spatial relationship in the following way: 'To be human is to live in a world that is filled with significant places: to be human is to have and to know *your* place' (p. 1, emphasis in text). The identification of significant places and the ownership of places also implies the construction of insignificant, low, and devalued places (Crow 1994). In part, this construction of devalued places is an outcome of the stigmatization of persons and groups. As Sibley (1995) argues, 'fear of difference is projected onto the objects and spaces comprising the home or locality which can be polluted by the presence of non-conforming people, activities or artefacts' (p. 91; also Lefebvre 1991).

The focus for this discussion lies in what Shields (1991) has called 'the logic of common spatial perceptions accepted in a culture' (p. 29), specifically the widespread perception of stigmatized places. How can we understand the integral and dynamic role of place in defining and reinforcing the social relations and practices of stigmatization? Shields (1991) provides a useful argument to frame the connections between stigmatization and place; his 'imaginary geographies' are expressions of classifications and divisions in society. But space is more than fantasies or metaphors for values, history, and culture; it is acted upon by social constructions, but also enables and constrains social activity. Anderson and Gale (1992) argue similarly, 'In constructing cultures, therefore, people construct geographies. They arrange spaces in distinctive ways; they fashion certain types of landscape, townscape and streetscape; they erect monuments and destroy others; they evaluate spaces and places and adapt them accordingly; they organize the relations between territories at a range of scales from the local to the international' (p. 4). In terms of homelessness and HIV/AIDS, the social construction of stigma is guided by the built environments associated with homelessness and HIV/AIDS and those designed to provide services for such individuals (Laws 1994).

Stigma and Place

If we begin with the premiss that stigma is socially constructed, we can argue that groups (such as homeless persons or people living with HIV/AIDS) are devalued in distinct ways given the nature of their stigmatization. As argued in Chapter 3, the wider public tends to understand homeless individuals and persons living with HIV/AIDS as being less productive, more dangerous, and more personally responsible for being (becoming or remaining) homeless or living with HIV/AIDS. But how does space constitute, constrain, and mediate such social relations? I explore the relationship between space and stigma with respect to homelessness and HIV/AIDS through three interrelated issues: (1) how the stigma associated with homelessness and HIV/AIDS is inherited by the landscape and particular places; (2) how individuals may become stigmatized through interactions with facilities and locales linked with homelessness and HIV/AIDS; and (3) how spatial constructions of stigma are constantly redefined to reinforce negative understandings of homelessness or HIV/AIDS or are appropriated to challenge existing characterizations.

Before I explore the person–place dynamic underlying the stigmatization of homelessness and HIV/AIDS, it is important to understand the constitution of place. There are multiple dimensions to place such as location, landscape, time, and personal interactions (e.g. home and community, privateness, rootedness, and even drudgery) (Relph 1976). The importance of place in understanding stigmatization lies not only in the location of physical buildings housing human services, but also includes the broader constitution of social relations in locations, that is, place as the setting for social interactions or locale (Agnew 1987; Goffman 1963; Thrift 1996). In terms of exploring stigma, however, place is more than just the setting for the stigmatization of persons, it also plays a fundamental role in representing and reproducing the marginalization of specific groups; as Pred (1985) argues, 'Place always represents a human product; it always involves an appropriation and transformation of space and nature that is inseparable from the reproduction and transformation of society in time and space' (p. 337; also Lake 1990). Sharon Zukin (1991) has described place as 'how a spatially connected group of people mediate the demands of cultural identity, state power, and capital accumulation' (p. 12).

There has been a substantial body of work especially in geography and architecture exploring the linkages between built form and social relations.[4] This research has indicated that built form includes the physical built environment, the cultural meanings and value expressed by and through this built environment, and the ways in which this environment is used (Cuff 1989; Goss 1988; Relph 1976). The linkages between spatial relations and marginalization have been particularly important in the exploration of gender and urban form. In terms of gendered urban form, two issues have been particularly important, suburban development and domestic labour (England 1991; Hayden 1984; Wekerle *et al.* 1980). These studies have indicated how the gendered division of labour within the household and its spatial implications in modern capitalist society have enabled and perpetuated the subordination of women. For example, in suburban neighbourhoods, individuals working full-time in domestic labour (usually women) are often isolated from a social support network of family and friends, and often must rely on public transportation systems designed for work trips. But marginalization and exclusion also take place on a building and even within-building scale. As Dolores Hayden (1995) has commented, '[a] territorial history based on limitations of gender in the public spaces of the city. . . would put buildings or parts of them off limits, rather than whole neighbourhoods' (p. 24). As she argues about twentieth-century spatial separation by gender, '[t]he segregation need not be absolute—women might be permitted to attend a class, but sit separately, or they might be allowed to enter a club as men's guests, provided they remained in a special room reserved for ladies, and so on' (p. 24). There are parallel discussions in terms of the racialization of space, from the association between race, class, and location in developing negative stereotypes concerning employability (Kirschenman and Neckerman 1991), to the linkage between crime, illicit drugs, and race in inner-city areas (Galster and Hill 1992).

Stigma and Locale

The definition of place as locale is central for understanding contemporary community response to controversial human services and their clients in its defining and reproducing notions of acceptability and non-acceptability (in persons and locations). Nigel Thrift (1996), following Giddens (1984) and the time-geography formulations of Hägerstrand, argues that there are five effects of dominant locales (or the locales where individual life paths must allocate time): (1) since life paths must flow through these locales, they become central in structuring individual life paths in class-specific ways through time and space; (2) since dominant locales constrain individual life paths, they also affect persons who interact with these individuals; (3) they are the sites of social interaction, where knowledge about the world is obtained; (4) they structure the daily routines defining individual activities; and (5) they constitute the primary site of socialization, 'within which collective modes of behaviour are constantly being negotiated and renegotiated, and rules are learned but also created' (p. 82).

In these ways, domiciled people on their way to work (to a dominant locale) walking or driving by homeless persons standing in line at a local mission are reinforced in their perceptions concerning the value of domiciled and productive places and practices, and the non-acceptability of homeless persons and locales. Homeless persons themselves also define their self-identity and those around them through the spatial confinements drawn by human service facilities and social control agents (such as the police), the opportunities for wage earning and obtaining resources, and the maintenance and search for dwelling sites (Rowe and Wolch 1990). The daily paths and routines of homeless persons are structured by their dominant locales (largely defined by the system of human service provision), and also serve to structure the stigmatization of homelessness. As Pred (1985) argues more generally, 'certain institutional projects are dominant in terms of the demands they make upon the limited time resources of the resident population and the influence they therefore exert upon what can be done and known' (p. 341).

Place Inherits Stigma

Stigmatization associated with homelessness and HIV/AIDS extends beyond the devaluation of individuals or groups. The landscape (i.e. the physical built environment, meanings and value, and activities in this environment) can also inherit the stigma associated with homelessness and HIV/AIDS, particularly with respect to the location of human service facilities. As such, as Sharon Zukin (1991) has argued, 'Themes of power, coercion, and collective resistance shape landscape as a social microcosm' (p. 19). The stigma attached to individuals and groups often becomes embodied in the human service facilities used and frequented by homeless persons and people living with HIV/AIDS. As Laws (1994) has argued, the built environment reflects the construction of stigma in society; thus a society which stigmatizes homelessness and HIV/AIDS will generally produce a stigmatized landscape.

If we extend the previous characterization of the social construction of stigma concerning homelessness and HIV/AIDS, then the embodiment of stigma results in facilities and their immediate vicinities also becoming defined as non-productive, dangerous, and evoking lack of personal responsibility. The stigmatization of places as non-productive emanates from social relations of production in similar ways to the stigmatization of persons and groups (Lefebvre 1991). As Agnew (1987) argues concerning the linkage between place valuation and production relations, 'By dint of living in a society in which the metric of economic transactions, exchange value, is the dominant measure of worth, people come to act it and use it as a natural measure of place-valuation and place-definition' (p. 34; also Logan and Molotch 1987). Thus, the distinction and spatial separation of normal communities from places associated with homelessness and HIV/AIDS are products of the social relations underlying production and reproduction, but also become the starting-point for further stigmatization of homelessness and HIV/AIDS. If the identity of a place is 'that which provides its individuality or distinction from other places and serves as the basis for its recognition as a separable entity' (Lynch 1960), then the association of places with non-productivity, dangerousness, and blame will generally result in the devalued identity of places associated with homelessness and HIV/AIDS.

Skid Rows: Stigmatizing Place Over Time

A stark example consists of the historical development of the 'place-myth' of Skid Rows in inner-city metropolitan areas (Shields 1991; also Davis 1990; Hayden 1995).[5] As an example of a reified spatial stigma, Skid Rows have become as much a part of mainstream society as they are a part of homeless existence (Anderson 1991). Skid Rows' physical built environment, everyday practices in the environment, and cultural meaning and value are in direct opposition to the ubiquitous social norms of nuclear family, the work ethic, and home-ownership. The processes of change in Skid Rows across the US have resulted not only in the material reality of concentrations of human services and service-dependent individuals, but also in their widespread understanding as places where the indigent and dangerous congregate (Davis 1990; Rowe and Wolch 1990; Ruddick 1996).

During the 1870s, Skid Rows developed in urban areas as businesses and charitable organizations began to provide for the needs of the transient labour force (Cohen and Sokolovsky 1989; Erickson and Wilhelm 1986). Local government and non-profit charitable organizations constructed settlement houses, and emergency and permanent shelters. In the early decades of the twentieth century, the expansion of the lumber and railroad industries, beginning prior to the Civil War, but escalating during the twentieth century, created a large demand for a 'large, ready, mobile, and single male labor force' (Vander Kooi 1973: 65). Skid Rows became centres for the transient labour market, with the consequence that demand for human services varied seasonally. In the winter months, when the demand for transient labour decreased, the need for services increased.

Residents of Skid Row areas were historically unique in their relatively ample employment and income opportunities, and their participation in political and union organizations (such as the Industrial Workers of the World). In New York City's Bowery in the 1930s, Skid Row men produced their own newspaper (The Hobo News) and organized an annual convention. In addition, the Hobo College emphasized activism and independence for those living and working in the area (Cohen and Sokolovsky 1989). During this time period, which Gilbert (1983) has termed 'welfare capitalism', Skid Rows also witnessed a significant influx of private charitable donations. The railroad industry, for example, without tax incentives contributed about a million dollars towards the construction and operation of YMCA buildings from 1882 to 1911 (Hall 1987).

From places where transient, seasonal labour was housed to contemporary locations where human services and their users congregate, Skid Rows have historically taken on the mantle of a 'zone of discard' or 'zone of dependence', building on Ernest Burgess's and the Chicago School's classic characterization of the 'zone of transition' (Blau 1992; Hoch and Slayton 1989; Dear and Wolch 1987). The Great Depression was a major turning-point for Skid Rows. The Depression marked an era of massive unemployment and dislocation, and, during this period, Skid Row populations were the largest in history (Hoch and Slayton 1989; Hopper 1990; Rossi 1989). During this period, housing policy began to shift from the private sector to non-profit and charitable organizations. Although the Second World War and its conclusion brought employment opportunities, with its expansion in the armed forces and defence-related industries, and a post-war economic boom, Skid Row residents were not participants in the expanding economy. Technological advancements in equipment radically changed the nature of skilled labour, significantly reducing the demand for a transient, predominantly unskilled labour force. The lack of seasonal and agricultural employment meant that transient men no longer dominated Skid Row populations; instead many residents were physically unable to work or not skilled enough to respond to the changing demands of employment.

The development of Skid Row as a zone of discard/dependence was facilitated in part by planning practices such as zoning and conditional use permits, which both blocked the siting of facilities elsewhere and eased the entry of facilities in Skid Row. Federal policies were also enacted through which the government could directly shape the development of downtowns (Mollenkopf 1983; Plotkin 1987). Urban renewal policies during the 1960s and 1970s cleared much of Skid Row and central downtown areas of flophouses, low-income neighbourhoods, and other signs of urban decay and deterioration, facilitating the division between Zukin's (1991) 'landscapes of consumption and devastation' (p. 5; also Davis 1990; Logan and Molotch 1987). While federal funding directly available to cities and neighbourhoods (such as the Community Development Block Grant and the Urban Development Action Grant programmes) brought development dollars, resulting in a gentrification of commercial and residential development, the outlying areas of downtowns became central locations for human services and shelters (Ruddick 1996).

Stigmatized and Privileged Places

The concentration of facilities in Skid Rows (and in places associated with high-density living conditions, low-income populations, and persons of colour) has created a spatial map of places which the public evaluates as relatively less productive, more dangerous, and personally culpable, while other places (such as residential and suburban neighbourhoods, workplaces, and shopping malls) are seen as acceptable and at times very desirable. Indeed, if home places constitute the 'foundation of our identity as individuals and as members of a community' (Relph 1976: 39), then the placement of facilities near our home places constitute vital threats to our own identities. Thus, the spatial separation and segregation of residential environments for homeless persons and people living with HIV/AIDS from wider society is an indicator and perpetuator of socially constructed relations of stigma. Stigmatized places make home and community more valuable in the hierarchy of spaces, especially if we view community as Young (1990) has defined it: 'a community is a group that shares a specific heritage, a common self-identification, a common culture and set of norms'

(p. 234; also Grant 1994; Mair 1986). Stigmatized groups are dispossessed of home (especially homeless persons) and community making them threatening symbols of anarchy, rebellion, or upheaval. The removal of such groups from the safety and predictability of home and community are thus paramount to maintaining order. Thus, Tuan (1979) argues, stigmatized persons are faced with exile or confinement: 'With exile, danger is expelled from the communal body; with confinement, it is isolated in space, thereby rendering it innocuous' (p. 187; also Plotkin 1987; Susser 1996). But such strategies also tend to exacerbate the stigmatization of homelessness and HIV/AIDS. The further stigmatized persons and places are from our home and community, whether the distancing is voluntary or involuntary, the more intense the fears of the unknown and the unfamiliar and potentially dangerous Other (Dear *et al.* 1997).

The existence of stigmatized places implies that home and community are free of the non-productivity, dangerousness, and personal culpability associated with homelessness and HIV/AIDS. Therefore, the construction and maintenance of abnormal, dangerous, and blamed places helps to maintain our imagined communities where life is normal, safe, and blameless (Anderson 1983; Logan and Molotch 1987; Massey 1994).[6] Indeed, wider society holds claim to the desired spaces of home and community, relegating homelessness and HIV/AIDS to lesser valued places. Community becomes synonymous with homogeneity, while difference is equated with danger and deviance: 'The ideal of community denies and represses social difference . . . the ideal of community denies difference in the form of temporal and spatial distancing that characterizes social process' (Young 1990: 227). The spatial expression indicated by the boundaries erected between home and stigmatized places has been described as a 'landscape of fear' (Tuan 1979). Landscapes of fear, largely associated with the city, have changed in their character since the eighteenth and nineteenth centuries, but have also retained remarkably similar themes according to Yi-fu Tuan (1979):

violent conflicts among urban magnates and the creation of a fortified landscape of fear; danger from and anxiety about strangers in an urban milieu; fear of anarchy and revolution, that is, of the overthrow of an established order by unassimilable and uncontrollable masses; distaste for and fear of the poor as a potential source of moral corruption and of disease; and urban fears in the lives of poor immigrants. (p. 157)

The spatial dispersion of homeless persons and people living with HIV/AIDS from the service-dependent ghettos of central city areas has unleashed this fear into the safe realms of home and community.

Human service facilities play primary roles in the definition and reinforcement of stigma since the stigma of places is conditioned in part by the built environment, the organization of institutions and groups providing services to homeless persons and people living with HIV/AIDS, and the daily/life paths of individuals inhabiting or passing through the space (Laws 1994; Rowe and Wolch 1990). Human services constitute powerful institutions in the lives of both the users of the facilities and wider society, since they structure the 'degree of fusion between local face-to-face social interaction and the integration occurring within and between the institutions of a social system' (Pred 1985: 340; also Snow and Anderson 1993). Just as neighbourhoods 'organize life chances in the same sense as do the more familiar dimensions of class and caste' (Logan and Molotch 1987: 19), the spatialized system of human services conditions societal and individual definitions of stigma.

Transferring Stigma From Places to Persons

The system of service providers acts directly to characterize or reinforce the stigmatization of individuals using the facility. There is a relationship of definition and reinforcement of stigma between the clients and the facility, so that persons using a homeless shelter are implicated in the stigma associated with homelessness (the same is true for facilities associated with HIV/AIDS). Service providers and the physical built environment they inhabit significantly control the environment, space, or locale where homeless persons and people living with HIV/AIDS interact and cope with everyday needs. The lack of control by homeless persons or people living with HIV/AIDS 'indexes one's social status and relative power within the community' (Rowe and Wolch 1990: 190), and illustrates the limited power that affected groups often have of changing characterizations of homelessness and HIV/AIDS. When dominant locales, such as Skid Rows, dictate the daily paths and routines of homeless persons through those locales, place reinforces the stigmatization of persons and persons reinforce the stigmatization of the place. As Laws (1994) has argued, 'human occupance of space can *reproduce* certain social values' (p. 1790, emphasis in text), and such environments both 'reflect and shape the identities of the residents' (p. 1791).

The transference of stigma from places to persons is not monolithic, but interacts with the characteristics of individuals. Stigmatization might be reinforced by use of a homeless facility for persons fitting a stereotypical image of homeless persons. Individuals not fitting the stereotypical image of homeless person or person with HIV/AIDS (whatever that image might be to the general public in any given time and place) might be better able to fit in with members of wider society, with minimal or temporary stigmatization presented by the facility. That is, individuals who do not reflect the public's stereotype of homeless person might not receive an immediate and permanent labelling as 'homeless person' given the use of a homeless facility. However, given the repeated use of a human service facility and specific individual characteristics, individuals might inherit the stigma associated with homelessness or HIV/AIDS from that facility. Such individuals become characterized as 'clients' or 'users' of the facility, implying their dependence on the human service delivery system and their subsequent makeover as 'homeless person' or 'person living with HIV/AIDS'. Since the built environment is relatively permanent, the use of human services or residence in a stigmatized place produces inertia in this stigmatization of clients, users, and residents.

The Dynamic Nature of Person–Place Stigma

But the stigmatization of places can undergo change, even with the relative permanence of the physical built environment. With respect to human service facility siting, such change has often meant the intensification of the negative stigma associated with individuals and location. Concentrating human services into specific areas of town, for example, tends to reinforce the stigmatized understanding of such areas. Examples include the identification of Skid Rows as harbouring homeless and indigent populations, or the linking of gay spatial enclaves with persons living with HIV/AIDS. The siting of facilities into specific areas reflects in part the historical planning practice of siting in areas of least resistance (i.e. places and communities not able or not willing to oppose such siting). Such policies exacerbate already unequal distributions of facilities, and, further, have the effect of intensifying the stigmatized character of specific communities. Thus, in similar ways that places can become racialized (Anderson 1991), the identity and widespread understanding of specific communities may become intertwined with the stigmatized identities of homelessness and HIV/AIDS. The location of human service facilities in specific communities means that the daily paths and routines of facility users centre on these particular facilities and their vicinities. Thus, as the place identity of that community becomes linked with human service facilities, the character of that community becomes devalued by wider society and even by residents of that community.

To prevent even the possibility of identification of communities as stigmatized through association with homelessness and HIV/AIDS, community residents may organize to prevent the proposed siting of human service facilities or to close existing facilities. In so doing, community members successfully use NIMBY tactics to maintain, enforce, and reinforce their boundary definitions, resulting in a maintenance of current spatial relations of stigma (Sibley 1995).

Circumventing and Challenging Stigma

Although the stigmatization of homelessness and HIV/AIDS implies a concomitantly stigmatized and devalued set of places, the person–place dynamic underlying stigmatization means that stigma may be circumvented and even directly challenged: 'The built and physical environments are negotiated realities, in other words, contingent outcomes of changing and often competing versions of reality and practice' (Anderson 1991: 28; also Berg and Kearns 1996; Davis 1990; Logan and Molotch 1987; Ruddick 1996). Consequently, the stigmatization of persons and places associated with homelessness and HIV/AIDS is not preordained, even with the vastly unequal power held by the 'unstigmatized.' Instead, the relations of socio-spatial stigmatization are also dependent on the changing identities and representations of homeless persons and people living with HIV/AIDS.

This is especially true in contemporary life, which some have described as a postmodern era. In this postmodern context, 'material existence . . . of embodied and emplaced processes of identity formation are being transformed' (Laws 1995: 256). The many facets of postmodern society which might contribute to the potential transformation in identity formation include the increasing attention paid to consumerism, surveillance, and the general instability in representations and identities. However, for homeless persons and people living with HIV/AIDS, the destruction, withering, and instability of widely held representations may result in various outcomes: circumventing the devaluation associated with homelessness and HIV/AIDS, experiencing intensified rejection and discipline by wider society, or directly challenging the stigma associated with homelessness and HIV/AIDS.

Circumventing Stigma

Circumventing the stigmatization associated with homelessness and HIV/AIDS has been an ongoing strategy used by affected groups to deflect devaluation and to facilitate fitting in to wider society, what Goffman (1963) has identified as 'passing' (i.e. concealment) or 'covering' (i.e. minimizing stigmatizing characteristics). That is, persons belonging to stigmatized groups make use of strategies which might reduce the social distance between themselves and wider society. As Jones *et al.* (1984) argue, '[t]he stigmatized, of course, are generally aware of the traits commonly associated with their social category and may frequently endeavor to avoid manifesting any of these traits' (p. 158; also Edgerton 1967; Scott 1972). Instead of being non-productive, homeless persons and people living with HIV/AIDS continue to participate in the paid labour force, while at times minimizing or obfuscating other characteristics which might emphasize their association with homelessness or HIV/AIDS. 'Hot bedding' (or taking eight-hour shifts for sleeping on beds in apartments), for example, is a strategy individuals have used to maintain the identity as homed and the material benefits of being domiciled (e.g. having an address, access to telephone and mail). Susan Ruddick (1996) argues further that 'invisibility is a means of access' (p. 57); as she describes, 'We are all familiar with stories of homeless women who pretend insanity to avoid attack, of people pretending to be traveling, watching movies, or waiting for a bus in order to find a sheltered place to sleep or sit' (p. 57).

Denial is also a familiar strategy to circumvent the stigma attached to homelessness and HIV/AIDS. Especially in relation to HIV/AIDS, the identification as living with HIV/AIDS brings up other contentious issues associated with transmission (e.g. homosexuality, intravenous or injecting drug use, and promiscuity). To prevent the assignation of stigmatization to self and family, individuals living with HIV/AIDS who are asymptomatic may not disclose their HIV-positive status to partners, families, and acquaintances, thereby ensuring their ongoing acceptance within their communities. Even in death, partners and family members may attribute illness and death to other terminal conditions (such as cancer) than associate themselves with HIV/AIDS.

The instability in identity formation also may contribute to an intensification of rejection among the populace with respect to homelessness and HIV/AIDS. The growth in post-industrial space in downtown areas, for example, has meant that persons seeking human services located in Skid Row must be removed for development to maintain its emerging value. Business groups have organized, for example, to remove homeless shelters in downtown Columbus, Ohio, because of potential value-depressing effects (Mair 1986).

Challenging Stigma

Homeless persons and people living with HIV/AIDS have also preserved these stigmatized identities in two ways: as a means of self-awareness and accessing public resources; and as political struggles to resist the stigmatizing characterizations which marginalize them and to redefine such stigma.[7] Bondi (1993) has argued that there are two aspects of such expressions of identity politics, one concerned with 'an emancipatory politics of opposition' and the other which is more individual, focused on self-discovery (p. 86). This second facet of identity politics, as Bondi (1993) describes, 'has the effect of replacing politics with therapy' (p. 86), and has been used as a way of reconnecting individuals with social support networks and as a means of accessing welfare state programmes and benefits. To reconnect individuals with welfare state benefits, for example, individuals must be identified as 'homeless' or 'living with HIV/AIDS' for eligibility. And thus,

the retaining of identity as 'homeless' or 'living with HIV/AIDS' is central
to maintaining access to public resources. However, this identification as
'homeless' or 'living with HIV/AIDS' is unlikely to result in emancipatory
or progressive action (Asch 1986).

However, identity politics have also been used in more emancipatory
ways. In such instances, 'the fiction of their coherency [e.g. the identity as
'homeless' and 'HIV/AIDS']' is necessary for political work (Fuss 1989:
105). Through the use of 'strategic essentialism' or Bell Hooks's (1990)
'choosing marginality', labels such as 'homeless' and 'HIV/AIDS' are being
appropriated by affected individuals and groups to change their meaning.
Identity politics have thus been used to create 'alternative designations of
persons subject to stigma' (Schur 1980: 192); 'AIDS patients' for example
are renamed 'persons living with HIV/AIDS' to influence the collective
consciousness. Members of the gay, lesbian, and bisexual communities have
been working to address directly both the prevention and stigma of HIV/
AIDS through their identities as 'gay', 'lesbian', or 'bisexual'. Although
homosexuality is largely associated by wider society with the transmission
of HIV/AIDS, Michael Brown (1995) has argued that 'gay men's bodies
and their spaces act *socially and contextually* to prevent the diffusion of
AIDS and HIV' (p. 169, emphasis in text), therefore, he continues, 'gay
spaces do not merely exhibit the diffusion of HIV, but also its blockade'
(p. 171). But choosing marginality is more extensive than the shifting of
labels. Soja and Hooper (1993) explore the decentredness of this strategy
in describing Bell Hooks's work, when they argue that 'she chooses a
space that is simultaneously central and marginal (and purely neither at
the same time), a difficult and risky place on the edge, in-between, filled
with contradictions and ambiguities, with perils but also with new
possibilities' (p. 190). In the material and identity decentredness of homelessness, homeless persons have worked to change the meaning of
homelessness, through art and photography exhibitions, the creation of
homeless writers' and artists' groups, and the development of newspapers
and publications by homeless persons (Ruddick 1996).

Conclusion

Stigmatization conditions and is conditioned by persons and places associated with homelessness and HIV/AIDS. The process of stigmatization is
thus inherently geographical. Homeless persons and people living with
HIV/AIDS, as perceived vectors of those conditions, experience individual
expressions of avoidance or disdain. Institutions also serve to reject
individuals characterized as homeless or living with HIV/AIDS through
local ordinances, such as anti-camping laws and zoning, and communities
organize to prevent the siting of human services which cater to their needs.

The socio-spatial stigmatization of homelessness and HIV/AIDS results in often intense conflict at the municipal and neighbourhood levels, as residents and local governments try to keep the deviance and difference associated with homelessness and HIV/AIDS from invading their communities and homes. Wider societal stigmatization of the persons and places associated with homelessness and HIV/AIDS reflects, and is manipulated by, these localized confrontations with stigmatized difference.

The stigmatization of homelessness and HIV/AIDS is far from being monolithic. Instead, community response to facilities and persons associated with homelessness and HIV/AIDS exists within a set of contexts: local and national as structural and institutional trends shape the current portrait of homelessness and HIV/AIDS; regional as proximate municipalities use protectionist and safety net ordinances to address homelessness and HIV/AIDS within their confines; and national as federal level policymakers, non-profit organizations, and interest groups work sometimes in direct opposition to deal with societal crises and to restructure the welfare state.

How specific communities respond to persons and places associated with homelessness and HIV/AIDS also depends greatly on their own positions in US society. As geographers concerned with community attitudes have emphasized, the characteristics of the potential host community (e.g. as potential hosts for human service facilities) are very related to eventual response to facilities associated with homelessness and HIV/AIDS. But position in US society is more broad than the socio-demographic characteristics of individuals and neighbourhoods. As researchers exploring the NIMBY syndrome as a facet of capitalist social relations have argued, class conflict and the process of production mean that controversial facility siting and hence community opposition may be inevitable components of production and development. But beyond class, there are two sets of relations which are central to the positioning of both neighbourhood residents and persons categorized as homeless or living with HIV/AIDS: race and gender. A comprehensive understanding of the changing nature of the NIMBY syndrome must address the intersection of these themes.

5

Race, Gender, and the NIMBY Syndrome

Much of the emphasis in the scholarly literature exploring the NIMBY syndrome has focused on localized opposition towards human service facility siting as an expression of exclusionary politics (whether explained as emanating from production relations or as a result of attitudes and behaviour). The rejection of human services is indeed in part traced to the stigmatization of persons and places associated with homelessness and HIV/AIDS (as indicated in Chapters 3 and 4), and the consequent desire of communities to keep homelessness and HIV/AIDS outside of their vicinities. However, the NIMBY syndrome is also more complex than solely exclusionary politics, with its implication of the marginalization of homeless persons and people living with HIV/AIDS by powerful individuals and groups, and the corollary of siting in marginalized low-income communities of colour. The formulation of the NIMBY syndrome as exclusionary politics benefiting the powerful is complicated by the complexities of race and gender.

Race has been defined in various ways, as biology, as ethnicity, as nation, and as social construction.[1] I use race in this chapter as a socially constructed set of meanings following Omi and Winant (1994) and Anderson (1991). In a conceptualization gaining wider scholarly appeal, Anderson (1991) has argued, '"Racial" differences cannot be conceptualized as absolute because genetic variation is continuous' (p. 12), therefore, demarcating groups by race has been arbitrary rather than expressing inherent essential biological distinctiveness. But even if race is not reflective of biological distinctiveness, it still carries powerful, popular influence, resulting in individual prejudice, institutionalized discrimination, and societal stigmatization. Omi and Winant (1994) in their widely cited book propose the definition of race as 'a concept which signifies and symbolizes social conflicts and interests by referring to different types of human bodies' (p. 55), arguing that 'we should think of race as an element of social structure rather than as an irregularity within it; we should see race as a dimension of human representation rather than an illusion' (p. 55).

Explorations of gender also stem from varying perspectives, but most search for explanations for the subordination and/or exploitation of women (e.g. Chan 1989). Feminist scholarship has increasingly critiqued totalizing feminist discourse, especially from perspectives such as lesbianism, race, and colonialism/post-colonialism (e.g. Collins 1990; de Lauretis 1990; Hooks 1990; Rich 1986). Gender and race must be considered in the context of homelessness and HIV/AIDS when exploring the constitution of the NIMBY syndrome, since these dimensions intertwine simultaneously to construct specific realities for persons homeless or living with HIV/AIDS, and for the communities confronted by possible human service facility siting.

In an effort to conceptualize the intersections among race, gender, and the NIMBY syndrome, this chapter focuses on two interrelated questions. The first is concerned with the racializing and gendering of homelessness and HIV/AIDS. The second centres on the relative importance of race and gender with respect to class, not only in community rejection of human services, but also in their unequal distribution in space.

The Absence of Race and Gender in NIMBY Research

Much of the research focusing on the NIMBY syndrome indicates that exclusionary land use politics is class based, enabling racism and retaining ethnic homogeneity (Plotkin 1987). Thus, there is a subtext throughout the literature on exclusionary locational politics that racism is at work. In critiques of populism, for example, scholars have outlined their suspicions of populist local politics which draw on nostalgic images of home and community to exclude and marginalize. For example, Cornel West (1986) argues: 'The weaknesses of populism consist of the worst of the xenophobic and jingoistic tradition of a European settler society: racism, sexism, homophobia, inward- and backward-looking, preoccupied with preserving old ways of life, defensive, provincial, and at times, conspiratorial' (p. 208). Researchers have noted, for example, racist subtexts in the language of local growth initiatives, gentrification, and downtown development, equating the inner city and particularly the black inner city as 'the wrong part of town, a place encompassing of otherness, whose spillover to other locations meant importing discord to more civilized city settings' (Wilson 1996: 84). But as West suggests, exclusionary land-use politics in general (and the NIMBY syndrome more specifically) do not emanate solely from oppressive social relations of production (although this is definitely a central element). In addition to production relations, the power relations of race and gender intertwine with homelessness and HIV/AIDS, creating a newly emerging portrait of stigmatization, marginalization, and, at times, challenges to the existing social order (following Massey 1994; also Omi and Winant 1994).

These issues with respect to controversial facilities and land uses have been taken up by researchers concerned with the problematic of race in land-use conflicts, especially in research exploring 'environmental racism' and the social movement termed environmental justice. Marginalization through the processes of environmental racism become expressed through inequitable exposure to environmental hazards by race and the mobilization of low-income communities of colour (e.g. Bullard 1990, 1993; Hurley 1995; Pulido 1996*a*). However, 'race' has not comprised a mainstay of research on the NIMBY syndrome directed against human service facilities. When race is discussed, it is often a side note, or a description of the characteristics of populations (e.g. those homeless or living with HIV/AIDS, or those who exhibit rejecting attitudes).

In general, scholars have noted that human service facilities have largely been concentrated in inner-city areas, particularly the Skid Row areas of metropolitan downtowns. Although there are human service facilities, even those associated with homelessness and HIV/AIDS, scattered throughout metropolitan areas, the most visible concentrations of facilities tend to be located in the historical service-dependent ghettos in older, central city areas. In considering these service-dependent ghettos as primarily areas inhabited by human services clients and low-income households, they have been largely associated with class (e.g. poverty and low socio-economic status), and less with race. However, as the discussions in this chapter and in Chapter 1 have indicated, homelessness and HIV/AIDS are becoming increasingly intertwined with race and gender.

While this implies that power relations inherent in race and gender play roles in the siting of human service facilities, there has been little exploration of how these roles impact the NIMBY syndrome. The exploration of race, for example, in the siting and opposition to human services cannot be seen simply as a 'monolithic racism' or 'racist' policies (following Pulido 1996*b*). Rather, the intersection of race with the NIMBY syndrome must be viewed in the context of homelessness and HIV/AIDS, in human service delivery, and in community response. If race is 'an unstable and "decentered" complex of social meanings constantly being transformed by political struggle' (Omi and Winant 1994: 55), then the intersection of race and the NIMBY syndrome has both exclusionary and empowering possibilities.

There is even less attention paid to the role of gender in the NIMBY syndrome directed towards human services, although this too is often implied in the literature on exclusionary land-use politics. As Sidney Plotkin (1987) argues, 'the prominence of neighborhood as a springboard of the new citizen activism is exemplified by the dramatic role of women as organizers and leaders of local opposition politics' (p. 33).[2] This has been especially true of participants in environmental justice social movements (Pulido 1998). Although researchers have argued about the central role of women in grass-roots activism, there has been scant attention paid to the gender relations inherent in the NIMBY syndrome, especially as it pertains to human service facility siting. The first step in understanding how race and gender play a role in the NIMBY syndrome is to explore how homelessness and HIV/AIDS are racialized and gendered. The second goal of this chapter is to outline the various ways in which the stigmatization of homelessness and HIV/AIDS has conditioned and been conditioned by the racializing and gendering of these two social crises.

Racializing and Gendering Homelessness

Race and gender differences have only recently become issues in the search for the nature, causes, and solutions to homelessness. The most stark gender differences over time emerge in comparisons made between contemporary homelessness and homelessness during the Great Depression in the US. Homelessness during the 1930s was mainly an issue for disaffiliated White males, often associated with seasonal employment opportunities and alcohol dependency (Hoch and Slayton 1989). These disaffiliated men, described as hobos or bums, were highly dependent on a complex network of social, housing, and employment resources historically located in Skid Row areas of downtown. The widely held image of homelessness was one of an ageing, alcoholic, White male with few social ties and at times needing public assistance. This mythologized image of homelessness continues to permeate the public consciousness today. Although the public's understanding of homelessness has changed particularly over the past two decades (with increasing visibility and diversity of the population, and growing scholarly and public policy attention paid to this phenomenon), pleas for charitable contributions (usually around Thanksgiving and the winter holidays) often draw on the mythologized image of the elderly, White, and alcoholic/mentally disabled male.

However, the increasing diversity of the contemporary homeless population has also become clear to the public, scholars, and policy-makers. The new homeless (in contrast to the old homeless of the Great Depression era) have been characterized as younger, better educated, more often consisting of women and families, having significant proportions of veterans, and consisting of greater numbers of persons of colour than in the past (Baker 1994; Bassuk 1991; Hoch and Slayton 1989; Rossi 1989). Although there is increasing diversity in the aggregate homeless population, there tends to be a set of geographic patterns in racial concentrations of homeless persons. Especially in inner-city, metropolitan areas, many studies of the sociodemographic characteristics of the homeless population have indicated the disproportionate representation of persons of colour, particularly African Americans, among those homeless and living on the streets (Bassuk and Rosenberg 1988; Takahashi and Wolch 1994).[3] The underrepresentation of Hispanics among those homeless is often attributed to the family-oriented social support networks buffering the structural, institutional, and individual challenges faced by this group (Baker 1994). The limited research on the linkages between race and homelessness is surprising given these often cited findings (Blasi 1994).

The racializing of homelessness is intricately linked to the changing nature of poverty and welfare in the US (discussed in Chapter 1). Concurrent with the changing demographic character of homelessness (that is, the growing proportion and concentration of African Americans), there has also been the added spectre of substance abuse (in particular, in inner-city areas, the growth of crack cocaine and other highly addictive substances). Injecting drug use, and ongoing policy debates about the appropriateness of needle sterilization programmes, make clear the potential for the criminalization of homelessness.[4] The association of homelessness with illicit drugs complicates community response, particularly among persons of colour, where community members both fear the identification as 'deviant' and the association with substance use, and at the same time constantly and vehemently search for solutions to this issue (Dalton 1989).

There have also been increasing numbers of women and their children counted among the new homeless, which Bassuk (1987) has termed the 'feminization of homelessness.' Women and men face similar challenges in negotiating life on the streets, including self-esteem/self-identity, acquiring daily resources, coping with illness and health risks, and seeking out and maintaining social support contacts (Bassuk 1993; Liebow 1993; Ritchey *et al.* 1991). Social support from friends, acquaintances, and lovers/spouses are particularly important in people's ability to cope with the social and material deprivation of homelessness. Lover/spousal relationships for heterosexual men who are homeless, in particular, may be very difficult to develop and maintain because of the large imbalance between the number of men and women living in places such as Skid Rows. Rowe and Wolch (1990) argue (for heterosexual individuals), 'even men who wish to enter into a lover/spouse relationship are constrained by a relatively small pool of available female peers' (p. 191).

But while there are similarities in the challenges faced by homeless men and women, studies have indicated that there are socio-demographic differences between men and women who are homeless. Homeless women tend to be younger and are more often persons of colour than homeless men, and, in addition, are more likely than men to have children with them (Baker 1994). Single homeless women also exhibit an almost 50 per cent higher rate than homeless men of hospitalization for mental disabilities (Burt 1992). Women's reasons for becoming homeless also differ from men's, for example, women tend to cite family disputes such as domestic violence as reasons for their homelessness (Marshall 1991).

Women who are homeless face particular challenges in coping with homelessness, especially the threat and fear of physical violence and maintaining custody of their children. Physical vulnerability conditions women's strategies in coping with day-to-day survival. Women who are homeless, for example, are twenty times more likely than women as a whole to be sexually assaulted (Wright and Weber 1987; also D'Ercole and Streuning 1990).[5] The risk of violence is not limited to life on the streets, but also occurs within the confines of emergency shelters, contributing to many homeless women's (and men's) decisions to avoid such housing options. Emergency shelters have also required invasive assessment and intake procedures, as a precursor to placement in a more appropriate facility (Golden 1992).

The influence of violence and the perception of physical vulnerability is evidenced by homeless women's development of social network ties. Strategies include entering into and maintaining relationships with men for protection, often even with men who are physically or emotionally abusive, joining street encampments (a strategy also used by single men and by couples), and trying to form personal and individualized relationships with non-profit and public service providers (Austerberry and Watson 1986; Goering *et al.* 1992; Rowe and Wolch 1990).

Women who are homeless and who act as the primary caregivers of children have additional issues with which to contend. On the one hand, women with children may encounter relative benefits when compared to men and women without children in terms of accessing public welfare benefits and social services (Baker 1994; Calsyn and Morse 1990; Susser 1996; Takahashi and Wolch 1994). Women with custody of their children are eligible for Aid to Families with Dependent Children (although such eligibility will change with recent federal legislation to reform the welfare system). Such benefits far outweigh the benefits available to individual men and women at the county and municipal level (through General Relief/Assistance benefits), thus those individuals (usually women) with custody of their children might be better able to obtain financial resources through existing welfare programmes. However, on the other hand, the benefits accrued to women through the custody of their children may be uneven and fleeting. The bureaucratization of dispensing welfare funds, and the instability in life paths and routines of homeless women (e.g. lack of personal connection with case workers, and personal crises such as robbery or physical assault), mean that acquisition of welfare funds is often haphazard. Landlords in rental housing markets are very much aware of these instabilities in financial resources and often refuse to rent apartments to women who have AFDC benefits (Rowe and Wolch 1990). In addition, women with children have needs beyond those of homeless women and men who do not have primary care responsibilities for children, such as day care and parenting skills training (DiBlasio and Belcher 1995).

Community Response to Homelessness

There is little research on the differences in community response by race and gender to human services associated with homelessness, except in the implication that communities which have opposed facilities associated with homelessness have been largely White, middle- and upper-class, and homeowners. In general, community response to homelessness is conditioned by relations of race and gender, specifically in terms of the characteristics of those homeless, and in terms of neighbourhood residents. In the research on community response to controversial human service facilities (most commonly associated with mental health care), women are often noted as being more accepting and less rejecting of facilities than men (Dear and Taylor 1982). In terms of race, anecdotal evidence indicates that persons of colour are no more accepting of human services associated with homelessness than White individuals, and indeed, exhibit similar tendencies of rejection. A national survey of attitudes towards a varied set of facilities indicated, for example, that African Americans do not have significantly different attitudes towards homeless shelters from White respondents (Takahashi 1997*b*). But this lack of variation in attitudes by race indicates that White middle- and upper-income communities may appear more oppositional because they have access to the mechanisms to express their opposition. That is, communities of colour may be no less rejecting of controversial human service facilities, but may lack the tools and resources necessary to prevent the siting of such facilities within their communities.

The nature of facility siting processes also plays a role in racial differences in opposition. Along with the structural and material limitations often encountered in communities of colour struggling with multiple issues such as poverty, violence, and unemployment, there is the added dimension that planners and policy-makers have historically seen such communities as sites of least resistance, where there has been limited vocal and effective opposition to siting. The concentrations of service facilities in Skid Rows and other inner-city metropolitan areas illustrate the effects of such strategies over time. Since policy-makers and service providers have looked to low-income, renting, communities of colour for siting multiple facilities, there have been relatively fewer instances of middle- and upper-income communities being targeted for facility siting. Thus, although many cases of the NIMBY syndrome are associated with White communities, one might guess that in the aggregate such communities might be disproportionately targeted less frequently as potential sites, and therefore, have less reason to express their opposition and rejection.

But when any community is targeted as a potential site for a facility associated with homelessness, public response may differ depending on the characteristics of homeless persons. Neighbourhood residents may react differently towards homeless men and women, particularly if the homeless women have children with them. In such instances, Baker (1994) argues that 'gender-typed definitions distinguish the "dependent" social construction typical of homeless women from the "deviant" more typical of homeless men' (p. 490; also Golden 1992).

In general, the stigmatization associated with homelessness by communities of colour interacts with the marginalization of those residents, resulting in a complex array of responses. When the NIMBY syndrome is practised by home-owning, suburban neighbourhoods, largely viewed as being comprised of White middle- and upper-income residents and households, community opposition towards facilities associated with homelessness is a statement not only of the deviance and stigmatization of homeless persons and places, but also as an expression of the solidification and practice of White identity. Thus, in successfully opposing human service facilities associated with homelessness, residents are also working to deflect efforts to breach the social, economic, and spatial boundaries defined as their communities. In maintaining a homeless-free environment, the dimensions of White urbanized identity as homed, suburban, and middle and upper income can be maintained. Further, inner-city problems such as substance abuse, violence (e.g. gang-related and domestic violence), property damage and deterioration (e.g. graffiti, abandoned housing) can be viewed as being prevented from entering the community when the communities' boundaries are re-articulated through practices such as the NIMBY syndrome. Not only may this practice help community members to maintain and reinforce internal social norms, but, further, any occurrence of these inner-city-type-problems (such as vandalism, graffiti, domestic violence) in the community may be understood as exceptions rather than indicators that the neighbourhood is deteriorating or on a downward slide. Hence, although not specifically expressed as racial projects, through successful NIMBY strategies to prevent the siting of human services associated with homelessness, residents often work to rearticulate White urban/suburban identity as explicitly devoid of homelessness (following Omi and Winant 1994).[6] That is, White urban/suburban identity is reinforced in its privilege (defined through productivity, lack of danger, and lack of personal blame) by spatially controlling the location of homelessness away from home, workplace, and recreational places.[7] Just as racism allows an imagined purity to be maintained, community opposition and the NIMBY syndrome help to maintain and reinforce communities free and pure of the material and moral deterioration associated with home-lessness (following Gilroy 1992).

But White communities are not the sole purveyors and practitioners of stigmatization through community opposition. Stigmatization of home-lessness may also become enhanced through the creation by subaltern or subordinated groups of their own Other (Laws 1994). In communities of colour, homeless persons who are of similar race may be stigmatized as being from outside the family/community, while those of a different race are viewed as being foreign, and their presence seen as an invasion. As Iris Young (1990) has argued, 'Members of culturally imperialized groups . . . themselves often exhibit symptoms of fear, aversion, or devaluation toward members of their own groups and other oppressed groups' (p. 147). With already marginalized places, communities of colour may act to protect what they perceive are dwindling amenities due to the siting of controver-sial human service facilities, especially when the users of the facilities are not members of the community (e.g. Pulido 1998). Thus, the opposition directed against human services associated with homelessness may be as or even more vehement than that encountered in largely White, home-owning, urban/suburban communities. The difference though may be in the material outcomes of such opposition. Communities of colour engaged in local mobilization over controversial facilities and land uses, although their visibility and credibility are increasing, still face the material obstacles of limited resources and access to powerful institutions and individuals that permeate many social movements initiated by low-income groups (Pulido 1996*a*).

Racializing and Gendering HIV/AIDS

Widely used epidemiological data continue to indicate (and construct) the racializing and gendering of HIV/AIDS. According to these data, HIV/AIDS has continued to be critical for men who have sex with men and intravenous/injecting drug users (IDUs); over eight in ten of the cumulative reported AIDS cases in the US have occurred among these two groups (CDC 1996; also Des Jarlais and Friedman 1994). These cases have tended to be spatially concentrated in the metropolitan areas of New York City, Los Angeles, San Francisco, Miami, and Washington, DC (accounting for nearly one-third of all cases nationally). However, while HIV/AIDS remains largely concentrated in White men through homosexual or bisex-ual contact, HIV/AIDS has also become increasingly problematic for women, persons of colour, and heterosexual individuals. In addition, HIV/AIDS has become significant outside the inner city, affecting even suburban residents (Orange County HIV Prevention Planning Committee 1995; Wallace *et al.* 1994).

As discussed in Chapter 1, the widely reported and accepted epidemiology of HIV/AIDS has indicated that African Americans and Hispanics are disproportionately represented among reported AIDS cases (this proportion is even greater for African American and Hispanic women), and for women the number of reported AIDS cases is growing faster than for men, with AIDS constituting the leading case of death for women in New York, New Jersey, and Puerto Rico (Amaro 1995; Diaz *et al.* 1993; Mays and Cochran 1988). For women, intravenous/injecting drug use and heterosexual transmission constitute the primary routes of transmission.

The disproportionate incidence of HIV/AIDS in communities of colour has motivated researchers to assess population knowledge and attitudes concerning HIV/AIDS, determine the effectiveness of public health strategies concerning HIV/AIDS education and prevention, and explore the practices and cultural norms defining sexual behaviour, particularly homosexuality (Cochran *et al.* 1991; Marin 1989, 1993; Morales and Bok 1992; Singer *et al.* 1990*a*).[8] Such research has identified varying knowledge gaps concerning HIV/AIDS among different communities of colour (although knowledge about HIV/AIDS has significantly increased over time in the African American population), outlined characteristics linked with differentiated knowledge (such as acculturation, income, and gender), and developed education and prevention strategies to control HIV/AIDS transmission (Van Vugt 1994). Studies have also moved beyond the documenting and classifying of HIV/AIDS cases in communities of colour to questions concerning the social construction of HIV/AIDS and the barriers to mobilizing around this issue (Singer *et al.* 1990*b*, 1991). Barriers to mobilization have been particularly acute for communities of colour as they grapple with the multiple stigmas of race, sexuality, and illness emanating from the public definition of HIV/AIDS particularly during the 1980s as a condition of White gay men (Quimby and Friedman 1989).

Researchers investigating the various dimensions of the social construction of HIV/AIDS have identified many of the mechanisms underlying the stigmatization of HIV/AIDS and its regulation. Social processes include the social distance between researchers and persons living with HIV/AIDS and the discourse surrounding health care and research (Brown 1995; Kearns 1996), the politics of sexuality, gender, and migration which define the epidemiology of HIV/AIDS (Patton 1990, 1994), and the local processes involving individual and community response to HIV/AIDS, often rejecting and oppositional in content (Takahashi and Dear 1997; but see Chiotti and Joseph 1995). Much of this discussion has centred on the gay/lesbian/bisexual dimensions of the stigmatization of HIV/AIDS. Indeed, geographers have been on the forefront of constructing geographies of HIV/AIDS which provide an alternative to the epidemiological stories often promulgated through the medical and public health literatures (e.g. Geltmaker 1992; Wilton 1996). While these alternative geographies are being increasingly developed, they still remain rare. Two important geographies requiring greater attention are the experiences of communities of colour and women, in particular, how the constitution of stigma surrounding HIV/AIDS is mediated and reflective of race and gender.

HIV/AIDS and Stigmatization: The Role of Race and Gender

The stigmatization of HIV/AIDS and its linkages with communities of colour are clearly indicated in the search for the origins of HIV/AIDS in Africa. As Cindy Patton (1990) argues, 'the best research minds of the Western world set off on a fantastic voyage in search of the source of AIDS. They went to Haiti and Zaire because the first non-Euro-American cases were diagnosed in people from these countries' (p. 83; also Dalton 1989; Treichler 1988). Although this African-origin hypothesis has become highly contentious, Africa remains a central focus for AIDS researchers primarily because of statistics indicating the disproportionate incidence of HIV/AIDS among Sub-Saharan Africans (Goldin 1994; Rushing 1995). In addition, popular publications have reinforced this racialization and geography of HIV/AIDS (e.g. Gould 1991).

The imagined invasion of HIV/AIDS from the African continent reflects the power of social construction in defining scientific research. Geographers, anthropologists, and other researchers have argued the centrality of the social construction of HIV/AIDS in understanding public response, scientific research, and public health policy (Brown 1995; Kearns 1996; Patton 1990, 1994). Such scholars have shown that the social construction of HIV/AIDS is framed by class, gender, sexuality, culture, and politics, in particular that HIV/AIDS has been socially defined as emanating from White gay men.

Although HIV/AIDS has been largely characterized and dealt with by the public, public health officials, and public policy-makers as a disease of White gay men, HIV/AIDS has also intricately become increasingly linked to structural relations defining race. Race of course cannot be divorced from relations of class, gender, and sexuality in exploring the geography of HIV/AIDS in communities of colour, and efforts have been made by a growing number of scholars to problematize race in conceptual understandings of HIV/AIDS (Wyatt 1991). There are two interrelated elements of this problematic: the social understanding of HIV/AIDS as a disease of gay men, particularly White gay men, and the lesser but still important connection between HIV/AIDS and intravenous/injecting drug use (Rogers *et al.* 1993). The epidemiological characterization of HIV/AIDS in the US as a disease of homosexual and bisexual men and injecting drug users has served to discourage engagement by communities of colour. In linking HIV/AIDS to an imagined community consisting of gay men and injecting drug users, the public was able to define boundaries serving both to reinforce their own safety and to delineate deviance and marginality from their own existence (Anderson 1983; Crimp 1992; Sibley 1995; Sontag 1988). Such boundaries blur for persons of colour in dealing with HIV/ AIDS as they struggle to determine their preferred group association and distance themselves from others.

HIV/AIDS as Homosexuality

The widespread understanding of HIV/AIDS as intricately linked to homo-sexuality has created significant obstacles both to public policies which deal directly with HIV/AIDS transmission and education, and to public acceptance of persons living with HIV/AIDS and facilities serving such populations. The threats to social norm structures and identity from homosexuality as mediated through HIV/AIDS crosses race, class, and even gender lines. As Young (1990) has commented, 'Homophobia is one of the deepest fears of difference precisely because the border between gay and straight is constructed as the most permeable' (p. 146). For persons of colour living with HIV/AIDS, the societal entanglement of HIV/AIDS with homosexuality (and substance abuse) often necessitates a choice between self-identifying as living with HIV/AIDS and possibly eliciting negative and rejecting responses, or living in silence and virtual invisibility. Patton (1990) has eloquently described this silence with respect to homo-sexuality and HIV/AIDS,

Silence: the never spoken, the yet to set itself into language, the unique, the individual, madness, the unrepresentable, the space of that which is not to be represented, the closet.

Silence: the unspeakable, the perceived but best not said, the ignored, the space occupied by that which is ignored, the hidden, the safely tucked away, the camouflaged, the safety of camouflage. (p. 129)

Silence and invisibility become particularly important for persons of colour living with HIV/AIDS since such choices become vital in determin-ing access to social and material support. Invisibility coupled with poverty and inadequate health care, for example, exacerbates the ability of persons of colour living with HIV/AIDS to cope with the physical and mental difficulties encountered after diagnosis with HIV/AIDS (Curtis and Patrick 1993; Roman 1993; Wilton 1996).

This silence for persons of colour living with HIV/AIDS has multiple facets. As HIV/AIDS has become socially defined as a condition of White gay men, the dominant representation of HIV/AIDS as White gay male has become widely recognized. Thus, as foundation dollars and public policy efforts become focused on HIV/AIDS education and prevention, much emphasis still remains tied to the White gay male population. This is not to say that such an emphasis is not necessary or completely misdirected. Epidemiological data continue to show that White gay males or White men engaging in same-sex or bisexual practices continue to constitute the largest group of reported AIDS cases in the US. However, the lack of equivalent recognition of persons of colour and women in the calculus of funding and programmes providing HIV/AIDS education, prevention, and treatment has been a central factor in the disproportionate representation of persons of colour among reported AIDS cases and the lack of medical diagnosis of AIDS-related gynaecological conditions for women living with and dying of HIV/AIDS (Amaro 1995; Patton 1994).

The solidification of the identity of HIV/AIDS as White gay male has had negative and positive effects on public response to HIV/AIDS, persons living with HIV/AIDS, and human service facilities. The identification of HIV/AIDS as significant for White gay men has resulted in the societal emphasis on homosexuality (rather than on race or class) in delineating deviance with respect to HIV/AIDS. This public understanding of HIV/AIDS as homosexuality has had individual, community, and societal impacts. At the individual level, people not identifying themselves publicly (or even to themselves) as homosexual or bisexual, even if they engage in same-sex or bisexual practices, are unlikely to be seen by service agencies as at risk, and, therefore, do not necessarily constitute a priority in terms of HIV/AIDS education and prevention services. Individuals who are living with HIV/AIDS, however, even if they are not homosexual are likely to encounter the societal stigma associated with homosexuality if they disclose their HIV-positive status. At the community level, the understanding of HIV/AIDS as homosexuality has meant that many communities are unwilling to have human services located nearby. But in contrast, places with activist gay, lesbian, and bisexual populations have worked to promote a positive gay identity, and in such places, human services for persons with HIV/AIDS have encountered acceptance and support (Chiotti and Joseph 1995). At the societal level, there are multiple interests working to promote acceptance for HIV/AIDS, and for homosexuality. The linkage between HIV/AIDS and homosexuality has been central to activism in the gay, lesbian, and bisexual communities for increased programmes and funding on research, treatment, and prevention. However, there have also been increasing incidents of violence against individuals perceived as gay or lesbian (Namaste 1996; Valentine 1993), and nationally documented examples of exclusion against school-age children, the most widely known being the case of Ryan White, where an HIV-positive youth was denied access to public school education.

HIV/AIDS as Intravenous/Injecting Drug Use

The complexities of HIV/AIDS and community response are also directly intertwined with the social relations underlying intravenous/injecting drug use (Bowser *et al.* 1990; Friedman *et al.* 1990). The social construction of HIV/AIDS as injecting drug use has located HIV/AIDS in the inner city and conjured images of criminals, deviant persons of colour, and violence (Quimby 1992). As Harlon Dalton (1989) comments about HIV/AIDS and drugs in the African American community:

We as a community have a complex relationship with illicit drugs, a relationship that often paralyzes us. On the one hand, blacks are scared to even admit the dimensions of the problem for fear that we will all be treated as junkies and our culture viewed as pathological. On the other hand, we desperately want to find solutions. For us, drug abuse is a curse far worse than you can imagine. Addicts prey on our neighborhoods, sell drugs to our children, steal our possessions, and rob us of hope. We despise them. We despise them because they hurt us and because they *are* us. (emphasis in text, p. 217)

For persons of colour not living with HIV/AIDS, the struggle for self-identification and escalating confrontation with HIV/AIDS often rests in class relations. As Quimby (1992) argues, 'Imagining HIV infection as afflicting poor Latinos and Blacks seems to permit the middle class a false sense of estrangement: Do we wait out the epidemic until it reduces the undesirable and disinherited among us or do we confront the issues that AIDS and HIV signify?' (p. 178; also Dalton 1989; Wallace *et al.* 1994). As Quimby implies in his statement, HIV/AIDS is not a condition experienced in isolation, but is grounded in the material existence of everyday life. The multiple challenges faced by communities of colour mean that HIV/AIDS often takes a back seat to more pressing issues. Poverty, crime, and unemployment are often more critical than the threat posed by HIV/AIDS. There is also the continuing suspicion, particularly among African Americans, that HIV/AIDS was created by the government 'as a means of wiping out black people' (Dalton 1989: 220; also Quimby 1993).

The lack of prioritization of HIV/AIDS, in combination with the linkages made between HIV/AIDS, homosexuality, and substance abuse, have contributed to the limited engagement by communities of colour with HIV/AIDS. When HIV/AIDS is confronted by community organizations, they must often tackle the interconnected problems of poverty, lack of access to health care, and discrimination which define broader life in these communities (Bracho de Carpio *et al.* 1990; Singer *et al.* 1990*b*, 1991). Newly arrived immigrant populations face the additional obstacles of limited English speaking or reading ability, lack of access to existing health care institutions, and mounting political and public pressures for withholding services.

Women and HIV/AIDS

And if homosexuals/bisexuals have been constructed as the vectors of disease for men in general (Brown 1995), the same could be said for women as the vector of transmission to their children, and the social construction of prostitution as the vector of HIV/AIDS for the heterosexual population. Female sex workers are at greater risk of HIV/AIDS because of their multiple sex partners and intravenous/injecting drug use, although such risk is tempered by class and access to resources.[9] However, the risk of HIV transmission to the heterosexual population posed by female sex workers remains highly questionable. As Fernando (1993) argues, 'The available evidence for the role of prostitutes as purveyors of HIV is meager and inconclusive' (p. 40). Even though female-to-male transmission has been argued to be much more difficult than male-to-male or male-to-female transmission, there remains the concern that female sex workers (and promiscuous women more generally) are a primary source of the spread of HIV/AIDS within the heterosexual population.

Gender also plays a critical role in the transmission of HIV/AIDS. The rise in HIV/AIDS among women through intravenous drug use and heterosexual contact reflect not only individual challenges concerning HIV/AIDS transmission (e.g. lack of knowledge concerning HIV/AIDS and behaviours placing individuals at greater risk), but also the social relations conditioning broader gender differences and inequalities (Amaro 1995). The prevention strategy to advise women and men to use condoms during sexual contacts, for example, does not take into account the unequal power relations within relationships and households between men and women, often related to economic dependence and cultural norms (Kline *et al.* 1992). Social norms concerning communication and sexuality play a central role in the ability of individuals to request the use of a condom, and, more generally, gender roles are integral in women's (and men's) abilities to carry out risk reduction strategies (Osmond *et al.* 1993).

The characterization of female prostitution and female sex workers as the vector of HIV/AIDS for the heterosexual population has meant a further marginalization of female sex workers, focusing their stigmatization not only on their depiction as deviant and dangerous, but now also as gateways of disease connecting the imagined geography of HIV/AIDS as homosexuality to the normal worlds inhabited by wider society. As a reflection of the depiction of female sex workers as the gateway between these two life worlds, efforts at HIV/AIDS education and prevention have often placed the responsibility for HIV/AIDS prevention on female sex workers rather than their partners/clients (Faugier and Cranfield 1995). Depictions of female sex workers as vectors of HIV/AIDS have implications for the ways that female sex workers (as vectors of HIV/AIDS for the heterosexual population) are socially and spatially managed. With its widening impact, HIV/AIDS has added a further marginalizing dimension to female (and male) sex workers, where female sex workers since the nineteenth century have already embodied societal boundaries between high and low social groups and places, constructing medical and social norm boundaries between normal and deviant (Hubbard forthcoming; Stallybrass and White 1986).

Conclusion

Conceptually, in Chapters 3, 4, and 5, I have argued for a more complex theory of the NIMBY syndrome. This theoretical framework will be illustrated throughout the subsequent chapters of the book. Using this perspective, I want to point out that I am operating from a set of underlying assumptions. The first is that the process of stigmatization should be central in any formulation of public response to homelessness and HIV/AIDS. It is through this process that abstract notions of difference become manifest in practices such as community opposition and exclusionary municipal ordinances.

We might consider the vitriolic response by a growing number of municipal governments when local homeless populations begin to expand, and cross some division (determined by the public and government officials) between an acceptable number of homeless persons and too many. The City of Santa Ana (in Orange County, California) for example in the early 1990s developed a municipal ordinance making camping outdoors illegal, primarily in an attempt to move homeless persons from the vicinity of the City's administrative centre (which also houses county and federal office buildings). It might be suggested that the development of exclusionary municipal codes, such as the one in Santa Ana, reflects the intolerant and rejecting attitudes of individuals (e.g. the workers in these government buildings) or the locational conflicts inherent in development and production relations (e.g. the reduction in potential business given the visibility and concentration of, in this case, homeless persons). While these are both undoubtedly central to local response, it is also plausible that these reactions by municipal governments (and by communities) reflect an additional premiss: the necessary benefits of boundary creation and maintenance through organized and institutionalized rejection of difference.

A second assumption follows from the first. Because the process of stigmatization is decentralized and dependent on individual, local, and regional experience with homelessness and HIV/AIDS, there are inevitably important local occurrences which result in a wide variety in public responses by place, time, race, and class (which are just four possible factors). There is an important degree of spatial distinctiveness within the stigmatization process, reflecting structural trends and local contingencies.

Thus, there is an inherent complexity within the stigmatization process which is central in understanding public response across time and space.

A third assumption is that public response to homelessness and HIV/ AIDS has both institutional and individual facets. That is, public response is not only comprised of individual community members organizing to resist the siting of specific human service facilities. Public response is also composed of the governmental institutions which respond to the changing visibility and societal importance of homelessness and HIV/ AIDS. Local-level institutions have the ability to respond to and incorporate state- and federal-level legislation and policies. Such institutions are by no means monolithic, but reflect the many interests of their members and constituencies. In other words, not all municipal agencies will respond in the same ways to homelessness and HIV/AIDS even within the same city or jurisdictional boundaries. Indeed, agencies often seem to act in contradictory ways, not as a co-ordinated, unified, internally rationalized organization. For example, while the City of Santa Ana was pursuing the development of the municipal code banning camping outdoors, it was at the same time working to develop a regional homelessness policy incorporating fair-share elements and human services co-ordination (perhaps not entirely contradictory policies, but seemingly distinct in their emphases).

In making the case for a more complex conceptualization of the NIMBY syndrome, I follow the lead of Robert Lake (1993) and Laura Pulido (1996a), in their investigation of environmental hazards, social movements, and institutional decision-making. They argue for greater contextualization and nuancing of theories concerning the placement of undesirable land uses, institutional motivations and practices leading to such placement, organized opposition and mobilization around such land-use controversies, and the influence and interaction of broader social relations, particularly of race and class, with locational conflict. Their perspectives make the NIMBY syndrome much more complex and provide a significant benchmark for understanding my formulation of this book. From case studies of media accounts of facility siting conflicts, to qualitative analyses of interviews with persons of colour concerning homelessness and HIV/ AIDS, and to institutional analyses of municipal and state responses, I argue that understanding the complexity underlying the NIMBY syndrome is based on an elaboration of the diversity of communities, homelessness, and HIV/AIDS.

Part III

Homelessness, HIV/AIDS, and Community Response

6

Rejecting Persons and Places

Locational Conflicts over Homelessness

It's sad, but I don't want to smell them.

'Tom', Screenwriter, late 20s, West Los Angeles (December 1996)

DURING the 1980s and 1990s, homeless persons and human service facilities experienced an increasingly vitriolic reaction from communities and municipalities. Wolch and Dear's (1993) study of community response to homelessness in Southern California indicates that even those cities which had in the past provided a more accepting and supportive environment were now encountering criticism of local policies providing services, public outcry over the growing and uneven distribution of homelessness (both individuals and social services), and an increasing tendency to turn to anti-homeless legislation to (re)move homeless persons from increasingly sensitive public spaces. The result of such trends has been that homeless persons have suffered a lack of accessible services and public marginalization with human services relocating or moving, and have encountered public demonstrations of increasingly vehement opposition. Enlarging services associated with homelessness in particular places or by particular organizations (predominantly in locations where facilities and services already exist), and legislative mandates requiring shelter (such as in New York) or income maintenance (such as in California) have rarely been adequate to respond to the growing need across metropolitan areas. Indeed, ongoing efforts at welfare state restructuring and welfare reform at the federal, state, and local levels have made the labour market the preferential societal mechanism for addressing those in need, even in the context of economic restructuring and its widely documented polarization of incomes and reduction in the number of well-paid manufacturing jobs.

If homelessness, and poverty more generally, were evenly distributed across the US, they might not elicit the public attention currently being placed on the creation of welfare reform legislation. However, homelessness is inherently geographical, varying in its significance and impact, with much of its potential effects to be experienced in cities and large

metropolitan areas where economies and histories are linked to immigration, the changing global division of labour, demographic change, and the
politics of human service provision. In these urban places, the negative
outcomes of economic restructuring and a repudiation of deviant
behaviours have become manifested as an increasingly visible population
of persons homeless (and/or living with HIV/AIDS), and a concomitant
compassion fatigue concerning the appropriate public response to social
crises.

Consequently, the rise of contemporary homelessness has acted as a
catalyst for explicitly expressing the stigma of persons and places by
individuals, community organizations, and local governments. *Ad hoc*
community organizations formed in response to a proposed or existing
human service facility, or to a visible group of homeless persons, constitute
one type of political response to homelessness. Municipal legislation and
ordinances marking non-acceptable locations for homeless individuals and
human service facilities, such as zoning ordinances prohibiting group
homes or municipal ordinances banning homeless activities such as panhandling or camping, are a different type of political response (and one to
which I will return in Chapter 8).

Although the US has witnessed a high degree of public rejection
directed towards homelessness, the stigmatization of homeless persons
and places has not been monolithic over time or over location. In
particular places, circumstances have contributed to the changing desirability of homeless persons over short time periods, while, in other
places, the stigmatization of homelessness has been more unchanging
and even further marginalizing. The local *ad hoc* politics of the stigmatization of persons and facilities is the focus of this chapter. I
explore these differentiated locational politics by analysing published
media accounts of the public response towards persons and facilities
associated with homelessness in Southern California during the 1990s.
Three instances of the NIMBY syndrome are presented: Skid Row, Los
Angeles; the community of Silver Lake in Los Angeles County; and the
northern county of Ventura, home to the city of Ventura (Fig. 6.1). At
issue is how we should understand the motivations and practices of the
individuals, organizations, and institutions involved in such conflict. By
outlining the diverse and at times changing stigmatization of homelessness, I hope to provide a concrete perspective on the themes raised
in Part II. To accomplish this, the chapter's objective is to place the
theoretical perspective on the stigmatization of persons and places
introduced in the previous three chapters in a specific context (i.e.
homelessness in Southern California in the early 1990s).

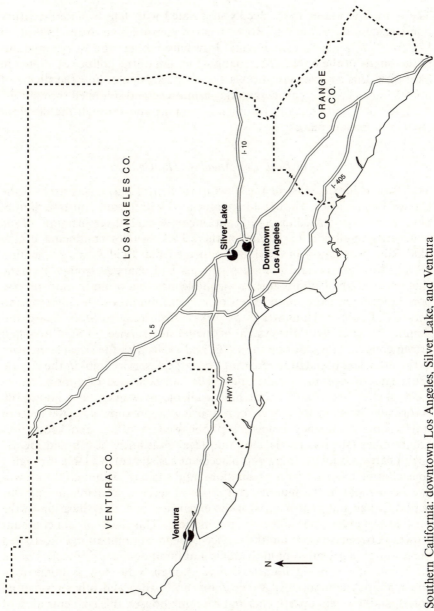

FIG 6.1. Southern California: downtown Los Angeles, Silver Lake, and Ventura

The Deepening of Stigma: Compacting Homelessness into Skid Row

There are few places more deeply associated with late twentieth-century homelessness than the Skid Rows of large metropolitan areas. Indeed, in Los Angeles, homelessness and Skid Row have become almost synonymous in the minds of the public. The changing nature of the politics of stigma in Skid Row, Los Angeles, constitutes a clear illustration of how the stigma of homelessness has become increasingly reinforced and solidified through the manipulation and control of homeless persons and through the development of homeless places.

Skid Row, Los Angeles: The Context

Skid Row is located east of the downtown central business district in the City of Los Angeles. Though long associated with a transient population, Skid Row experienced a significant increase and diversification of its homeless population beginning in the 1970s. Studies conducted of the Skid Row population in the 1980s indicated that Skid Row's population of very low-income and homeless persons had changed from the classic characterization of White, elderly, alcoholic men to a younger, more diverse population, consisting more often of women, families with children, and persons of colour (Hamilton *et al*. 1987). According to Stacy Rowe and Jennifer Wolch (1990), the system of shelter and services in Skid Row had grown concomitantly during the 1970s and 1980s with the sizeable increase in the homeless population, creating the largest 'service hub' in the city (p. 186). This service hub by the late 1980s had consisted of approximately 2,000 shelter beds (with about half available to women, and about 100 exclusively for women), longer-term housing (approximately 6,700 units in single room occupancy hotels), a public welfare office, and about fifty programmes (such as meals, clothing, legal assistance, health and mental health care, and job training and placement assistance) available through a constellation of social services, missions, and shelters. Although the growth had been rapid in the number and types of services available during this period, it had not kept pace with the expansion and diversification in the very low-income and homeless population. This lack of concomitant growth between services and the service-dependent population exacerbated the resource deprivation of individuals and families already homeless or on the brink of becoming homeless. And though Skid Row is home to a housing stock composed of single room occupancy and other residential hotels, and low-rent apartments and rooming houses, this concentration of housing was severely depleted by demolitions during the 1970s and 1980s. For example, the number of single room occupancy units fell by more than 2,000 between 1969 and 1986 (Hamilton *et al*. 1987).

The history of Skid Row, its residents, the changing roster of service providers and religious organizations, politicians, and business owners portrays a long-term struggle over the power and influence associated with Downtown development (Davis 1991; Ruddick 1996), and recent conflicts over the appropriate location of homeless services/shelters exemplify the ongoing solidification of the stigmatization of homelessness. Such conflicts express the complexity of the NIMBY syndrome as it relates to the location of homeless persons and places. In the first instance, that of relocating the Union Rescue Mission to a site more centrally in Skid Row, the City's participation in managing homelessness through concentrating services in Skid Row seems clear. The second example, that of a proposed drop-in centre for homeless persons on the fringe of Skid Row, would also appear to express NIMBY sentiments on the part of City and business interests. However, in both these examples, homeless advocates and service providers were in conflict about the acceptability of these facilities/centres in the service hub of Skid Row. The fragmentation of political views resulted in some advocates and service providers in opposition to the siting of larger, newer facilities for homeless persons in their vicinities, often citing similar reasons for more common NIMBY arguments (e.g. facility saturation, increased crime rates). Thus, the politics of homeless relocation did not always pit the City and business interests against homeless advocates and service providers in a clear struggle with clear and monolithic motives. Rather, these two examples in Skid Row indicate how the proposal and siting of facilities serving homeless persons may in themselves express the deepening of stigma, while the opposition to facilities may indicate a battle to reduce stigma and marginalization.

The Relocation of the Union Rescue Mission

The Union Rescue Mission has, for seven decades, provided food and religion to needy individuals, originally as a gospel food wagon and then later as a religious mission. Begun in 1891 by Lyman Steward, then president of Union Oil Company, the Union Rescue Mission began operations at Main and 2nd Streets in 1926. During the late 1980s, the City's redevelopment authorities encouraged the Union Rescue Mission to move from its Downtown location on Main and 2nd Streets to Skid Row's 'epicenter' (Gordon 1994). This move by City officials reinforced informal suspicions long held by service providers involved with homelessness, that City officials were working to relocate homeless persons and facilities providing services to the homeless population from the Civic Center area further east towards the central core of Skid Row. City redevelopment officials argued that services would improve when the Mission moved to Skid Row's geographical centre (Fig. 6.2).

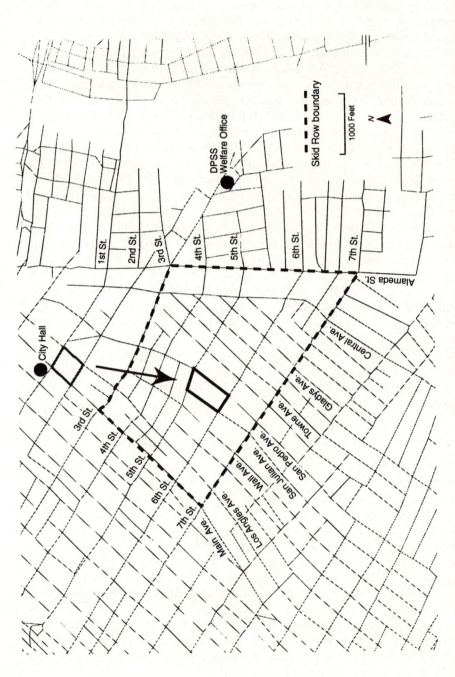

Fig 6.2. Skid Row, Los Angeles: the relocation of the Union Rescue Mission (adapted from Rowe and Wolch 1990)

The City since the mid-1980s had begun a strategy to rehabilitate the neighbourhood surrounding Union Rescue's Main and 2nd Street site, which was located near St Vibiana's, a centre for Southern California's Catholic archdiocese. As part of this strategy, City officials approved the construction of a now-open state office building (named for former California governor and US president Ronald Reagan) a block away from the cathedral. The City's Downtown Strategic Plan outlined a rehabilitation of the Mission's site through the demolition of the Mission and the construction of a new pedestrian plaza, low-rise housing, and commercial buildings directly south of St Vibiana's cathedral, dubbed 'Plaza St Vibiana' (Gordon 1994).

As part of the incentive to move, the Community Redevelopment Agency (CRA) had originally offered Union Rescue an agreement whereby the CRA would acquire a parcel of land through eminent domain and give it to the Mission, and, in return, the Mission would sign an agreement that, in its hiring of Mission staff, it would not discriminate by religion (Gordon 1994). Union Rescue officials did not agree to this, as they felt this would hinder their religious mission, central to their philosophy of service provision.

The CRA continued to pursue relocation of Union Rescue and in 1989 decided to approve a $6.5 million relocation grant (from public funds) to allow the Mission to purchase the land itself. In addition, the CRA planned to purchase the existing property on Main and 2nd Street for about $1.5 million. This agreement allowed the Mission to implement its own hiring practices, dropping the requirement for non-discrimination in employment by religion. The new Mission, with entrances on San Julian and San Pedro Streets, would be three times the size of the old building, at an estimated cost of $29 million. Along with the funding provided through CRA, the remaining construction costs were funded through donations and bequests, including a $11 million anonymous contribution and a $3 million contribution from Louis Foster, the founder of 20th Century Insurance, and for whom the new building would be named (Gordon 1994).

The design, construction, and operation of the new Union Rescue Mission were not heartily accepted by Skid Row businesses and shelter and service providers, however. Business owners on San Pedro Street, for example, were extremely apprehensive about the relocation of the Mission, and fought to prevent its siting in its new selected location. Under pressure, the CRA worked to have the entrances redesigned so that the main entrance to the Mission would be on San Julian Street (for meals, showers, or beds), while the San Pedro entrance would serve Mission staff and longer-term residents. In addition, a landscaped courtyard and indoor recreation rooms were designed to address the concerns of hotel operators in the areas, who believed that there would be massive line-ups and large crowds on San Julian waiting for Mission drop-in services (i.e. meals, showers, and beds).

Other service providers in the area were also worried about the crowding and potential increases in crime on San Julian with the opening of the Mission. As Andy Raubeson, executive director of SRO Housing Corporation, commented, 'The neighborhood is very apprehensive' (Gordon 1994: B1). Other service providers were angered by the approved design alterations, particularly the construction of dual entries into the Mission complex. As Alice Callaghan, executive director of Las Familias del Pueblo, argued, 'It's amazing to me that the concerns of businesses to have as few poor people on San Pedro as possible was the highest priority' (Gordon 1994: B1). Nevertheless, despite the concerns and arguments of various parties, the Union Rescue Mission opened its new facility in the autumn of 1994. The five-storey structure consisted of 235,000 square feet, holding up to 810 beds and capable of serving 2,000 meals a day. Warren Currie, the president of Union Rescue Mission, compared the two Missions: 'Our old facility was a one-lane road to help lead people out of Skid Row. Our new one is a three-lane major highway' (Gordon 1994: B1).

The 'Drop-In Centre' Controversy

The relocation of the Union Rescue Mission did not completely remove homeless persons from Downtown, nor did it resolve business concerns about homeless persons camping in open space and hanging around on nearby sidewalks. Business concerns stemmed from the potential or actual effects of the presence of homeless persons on customers, or at times emanated from outside sources, such as regulatory agencies. As one worker in Downtown argued, 'Downtown is too accommodating. There are too many bars and missions down here' (Williams 1994: B3). Such views often translated into business complaints to the Los Angeles Police Department, which would then move homeless persons from specific streets and confiscate personal property. Businesses also called on police to relocate homeless persons and belongings as a consequence of regulatory actions and warnings aimed at food industries. In November 1994, for example, Tengu Beef Jerky Company, located at Fifth and Towne Streets, called LAPD officers to clear the sidewalks in front of the business. This sweep occurred as a consequence of US Department of Agriculture warnings to keep Tengu's sidewalks clear of trash or waste. Michi Osaki, Tengu's office supervisor, argued that 'her company wants to be sensitive to the needs of the homeless but had no choice other than calling upon the city for help' (Daunt 1994c: B1). As the LAPD's Central Division's homeless coordinator, Sgt. Robert Veliz, argued, 'Because of health reasons, we have to respond to the complaints. [The businesses] have a legitimate concern. They have to deal with the transients and to be written up by the USDA. Why should they [the businesses] stay in Downtown Los Angeles?' (Daunt 1994c: B1). Such police sweeps, however, were seen by homeless service providers and homeless persons themselves as part of the ongoing strategy by the City and the LAPD to move the homeless population to a less visible location where routine business routines would not be disrupted.

Police sweeps and confiscation of personal property have often been used when conflicts flare over the presence of homeless persons. Downtown business interests during the early 1990s felt especially affected by an increasing homeless population. The Central City Association, representing business interests, characterized the sweeps as a necessary response to potential health and safety risks posed by homeless persons to businesses and consumers. Carol Schatz, senior vice president of Central City Association, argued that 'It is an issue of balancing rights. If [homeless persons] are affecting the health and safety of the people who purchase products from the packing plants, then something has to be done' (Daunt 1994c: B1). In November 1994, the City responded by using police sweeps affecting 30–40 per cent of homeless encampments in Skid Row (Daunt 1994c). The City's arguments for such sweeps centred on the necessity of creating and maintaining a 'healthy, business-friendly environment' (Daunt 1994c: B1).

But sweeps are at best only a temporary measure for addressing the local homeless population in Skid Row. With no permanent way of addressing the reasons for homelessness, police sweeps and confiscation of property only serve to relocate homeless persons temporarily to locations not visible to complainants. But without services, shelter/housing, employment, and other wage-earning potentials, homeless persons will be forced to return to places and sites whereby daily survival can be managed.

Perhaps in recognition of some of these more intractable issues, in the autumn of 1994, the City planned a drop-in centre, to be funded by 20 per cent of a $20 million US Housing and Urban Development grant to the City. This drop-in centre was described by City officials as a 'fenced-in urban campground where up to 800 people could take showers and sleep on a lawn' (Dear and Wolch 1994: B7). The plan from Mayor Richard Riordan's office would have the City purchase a vacant lot east of Downtown (i.e. in Skid Row) in a secluded industrial zone. The drop-in centre would primarily consist of a fenced lawn (approximately 80 per cent of the site), together with a 50-bed shelter offering showers and a centre for social services. The Mayor's plan envisaged a centre serving approximately 800 persons per day. Homeless persons would be shuttled to the site in vans staffed by social workers. As a joint City–County endeavour, the centre had to be approved both by the City Council (after review by two City Council panels, the Community and Economic Development Committee and the Community Redevelopment Committee) and by the County Board of Supervisors.

Homeless advocates and service providers reacted immediately to oppose the development of this drop-in centre/camping area. Advocates argued that this centre would become a large-scale urban campground, that it would encourage criminal behaviour, and that it would cause opposition from local businesses and Central City East neighbours (Dear and Wolch 1994). The fencing of the site and the shuttling of homeless persons to the centre were also troubling to homeless advocates. Gary Blasi, member of the Los Angeles Coalition to End Homelessness and a law professor at the University of California-Los Angeles, argued that 'We see it as a first step on a slippery slope to concentration camps in rural areas for homeless people. No one wants to take that first step. It's dealing with homelessness as an aesthetic problem for the Downtown business community' (Daunt 1994*a*: B1). Gene Boutiller, executive director of the Los Angeles Homeless Services Authority (a City–County agency charged with development and implementation of homeless policies), however, defended the Mayor's plan, arguing that the centre would provide crucial services and that 'if I build a park and you chose to lie down and take a nap, and you have other alternatives, I think that's fine' (Daunt 1994*a*: B1).

The Mayor's office met with homeless advocates in a closed-door meeting after the public outcry over the plan, and adjusted the proposal from an 'urban encampment' to 'service/resting area' (Daunt 1994*a*: B1). Rae James, Deputy Mayor, after this meeting stated that this centre would not be a campground, and no overnight sleeping would be allowed. After these changes to the plan, the Los Angeles City Council voted 13–0 to approve the $4 million centre, arguing that 'the city of Los Angeles wants to give people a ladder to climb out of the situation of homelessness' (Mike Hernandez, City Councilman), and '[the drop-in centre] is a first step out of homelessness' (Mark Ridley-Thomas, City Councilman) (Daunt 1994*b*: B1).

Although the proposal appeared to have wide approval from local elected officials, by May 1997 the proposal had still not been implemented. The modified service/resting area 'is over budget, two years behind schedule and in danger of still more delays' (Newton 1997: A14). But while the service/resting area remains mired in bureaucratic obstacles, Mayor Riordan, just at the beginning of his second term, is working to design and implement an anti-panhandling ordinance, a promise from his first term campaign, which 'would make it illegal to beg aggressively for money' (Newton 1997: A15). Riordan argues that such an ordinance is necessary not only to respond to business complaints about the ever-expanding homeless population, but also because '[i]t doesn't help the homeless to have abusive panhandlers roaming the streets, many of them committing crimes' (Newton 1997: A15).

The Ambiguity of Stigma: Parking your Van in Silver Lake

While the public's image of a homeless person often centres on the populace inhabiting places like Skid Row, Los Angeles, there are other groups of persons who may also become captured by the stigma attached to homelessness. Such groups become associated with all the negative characteristics assigned to the monolithic group labelled 'homeless', including mental disability, alcohol and substance abuse, criminality, and even perversion (Gallagher 1994). In Silver Lake, California, located north of downtown Los Angeles, a group of persons living in their vehicles became the focus of neighbourhood ire after residents realized that these individuals were homeless.

Silver Lake: Diversity North of Downtown

Silver Lake is a community located in the hills north of downtown Los Angeles. The community is bounded on three sides by freeways emanating outward from downtown (the Golden State, the Glendale, and the Hollywood freeways) and overlooks Silver Lake Reservoir. In the early 1980s, the neighbourhood was described by a resident as 'the kind of area for people who want to live a middle-class life but for one reason or another can't quite manage it in a traditional neighbourhood' (Dreyfuss 1984: V1). As a haven for individuals not fitting a traditional image, Silver Lake became home during the 1980s to a diverse population of residents, including Asians, Latinos, interracial couples, gay and lesbian households. The neighbourhood has remained a desirable location because of its proximity to downtown and Hollywood. A *Los Angeles Times* reporter commented in 1984 that 'Those who are uncomfortable outside the mainstream would probably not be comfortable living there. But almost anyone can find something in Silver Lake worth visiting' (Dreyfuss 1984: V12).

Since the 1970s and 1980s, Silver Lake has witnessed opposing trends of White flight and increasing racial minority populations, and gentrification with its concomitant growth in higher-income households, many of whom are politically active gay and lesbian individuals. Such diversity in the community has at times led to conflict and violence. The community during the 1980s experienced an increasing presence of gang activity. In 1992 there were about a dozen reported violent assaults directed at gay individuals, including beatings by youths with baseball bats. About eighty-five gay and lesbian residents marched to protest against these crimes, calling for increased police protection and the use of neighbourhood patrols (Kazmin 1992). The community has addressed the growth in potential conflict in innovative ways. In the late 1980s, for example, various community organizations (including the Sunset Junction Neighborhood Alliance, an organization associated with gay residents, and El Centro del Pueblo and the Central City Action Committee, organizations working with youth and gangs) worked together to initiate the Sunset Junction Street Fair.

In the context of a diverse and at times contentious neighbourhood environment, homelessness has not been a high visibility issue. Initially, the homeless individuals living in Silver Lake in the early 1990s did not experience a backlash from local residents while they remained unnoticed. These particular homeless persons did not necessarily fit the stereotypical image of persons living on the streets. Thus, when residents in the neighbourhood began to notice unfamiliar vehicles parked on their streets, such an occurrence probably did not elicit complaints to local government officials or police departments.

Unfamiliar Vehicles and Homelessness

The presence of unfamiliar vehicles may not spark concern among local residents. Such vehicles may constitute legitimate and temporary visitors to the community, friends, family, or acquaintances of home-owning or renting individuals residing in the vicinity. However, when such unfamiliar vehicles remain on specific residential streets for longer than some appropriate period of time, residents may begin to suspect that such individuals actually are not visiting residents living in the neighbourhood, but more likely are individuals living in their vehicles. Such a realization and the subsequent identification of vehicle owners by neighbourhood residents as homeless may then elicit responses to social control agents, such as the police or local government officials.

Such was the case in Silver Lake. In the autumn of 1994, residents in Silver Lake began to complain to their City Councilwoman, Jackie Goldberg (the first openly lesbian woman to be elected to local office) about individuals living in trailers, vans, and other vehicles which were parked on residential streets. Many of these individuals were not employed, none of them had telephones, and many used showers and sinks at the local swimming pool (Ni 1994). These individuals, subsequently identified as homeless and transients, were accused of urinating, defecating, littering, panhandling, and intimidating home-owning residents. There had also been a series of car burglaries and graffiti in the neighbourhood which were attributed to the individuals living in their vehicles. Residents complained that transients would park all along Riverside Drive (a major thoroughfare in the community).

Councilwoman Goldberg's office, in response to the resident complaints, began a regular garbage pick-up service near the site where the vehicles were parked, placed garbage cans and a portable toilet near the vehicle parking site, and restricted parking for these vehicles to a one-block area (Ni 1994). One of the home-owners, John Ruege, who had initiated and maintained opposition to the location of the vehicles and their inhabitants, argued that these responses were inadequate to resolve the problem; he argued that 'I know it's [Goldberg's responses] part of an effort to clean up. But that's also welcoming them. It's telling them it's OK to live there. There needs to be a real solution' (Ni 1994: B1). That is, Ruege's argument was that the vehicles and their inhabitants did not belong in a residential neighbourhood. Ruege argued, 'A city street is not a trailer park. Get the people off the street and find them jobs' (Ni 1994: B1). A local neighbourhood organization formed to respond to the unresolved controversy concerning the parked vehicles. Michael Frances, a representative of the Riverside Drive Neighbors' Alliance, argued in a common theme used by NIMBY proponents that the presence of the vehicles lowered property values and depressed the rental market.

Trailer and van residents, however, believed that they were being singled out for negative occurrences in the neighbourhood. Stan Carpenter, living in his trailer, argued that the people living in their vehicles worked to keep the area clean, and also called Councilwoman Goldberg's office if they saw anything 'unusual' happening in the vicinity. In addition, Ralph Smith, who lived in a Dodge van, argued concerning the rash of car burglaries, 'Do you think if somebody is going to be living somewhere seven days a week, with their license plates exposed, they would wander off right in their neighborhood and rob somebody?' (Ni 1994: B1).

The conflict in definitions over ownership and belonging to the community are clear. From the point of view of home-owning residents, the individuals living in their trailers represented a clear threat to their property (in terms of property values and material damage and theft). The residents of the varying types of vehicles saw the neighbourhood as theirs as well, with their identity wrapped up not in a permanent structure (i.e. a house) but in their identifiable vehicles (i.e. with a licence plate). Many of these vehicle residents had no intention of living on the streets, but enjoyed living in motor homes or vans, and saw themselves as protectors of the neighbourhood. A member of Councilwoman Goldberg's staff, Connie Farfan, supported this view, stating that 'In the last four to five months we haven't had more than three to four [telephone] calls. I don't think it's as disruptive to the community as a few people say it is' (Ni 1994: B1).

Even so, Farfan developed a proposal to develop a trailer park as part of the drop-in centre in Skid Row (the 'urban campground' proposal which was creating conflict between City staff and homeless advocates). If the trailer park proposal were accepted by the City, then the vehicles parked in Silver Lake would be forced to abandon their current location and moved to the site east of downtown Los Angeles. Gene Boutillier, executive director of the Los Angeles Homeless Services Authority, supported the idea of a trailer park in principle: 'That's a strategy that ought to be part of the mix of alternatives' (Ni 1994: B1). While City officials decided their fate, the fleet of vehicles remained restricted to one block located about a block away from nearby residences on Riverside Drive 'with the Golden State Freeway on one side and a steep embankment on the other side' (Ni 1994: B1). The proposal to resolve Silver Lake's homelessness issue using the urban campground in Skid Row indicated that the small scale of homelessness in Silver Lake was not seen by residents or local policy-makers as isolated from the larger Los Angeles metropolitan area, and thus, when considering options about how to address conflict, policy-makers such as Connie Farfan, expressed interest in the centralized solutions proposed by other Los Angeles politicians and policy-makers.

The Dynamism of Stigma: Ventura River Bed Inhabitants

Riverbeds, along with highway off- and on-ramps and other natural and built structures, are providing increasingly used alternatives to abandoned buildings, shelters, and street corners for the expanding and diversifying homeless population in the US. Although increasingly popular, riverbeds in large urban areas are not upscale havens, being rather sites often inundated by chemicals and waste, where hazardous wastes and garbage are illegally dumped, and where flash floods can occur without warning. Even with these dangers, homeless persons have found such sites increasingly preferable to places monitored by social control agents such as police, and, in Southern California, Immigration and Naturalization Service authorities. In addition, homeless persons often deem such sites to be much safer than places such as Skid Rows, which tend to be places of 'spontaneous violence' (Sahagun 1990: A1). As one homeless Latino man argued, 'Here, nobody gets arrested, beat up; nobody sells drugs or begs for money. The people here are peaceful' (Sahagun 1990: A3).

Riverbeds are becoming sites of increasing contention as public health authorities, Army Corps of Engineers, and environmental protection and local flood control agencies note the growing numbers of individuals living in riverbeds and using river water for daily living. Residents have also increasingly complained to government authorities about persons living in riverbeds, arguing that riverbed persons have harassed users of nearby parks and trails (such as joggers, bicyclists, and horseback riders) (Sahagun 1987a). For the most part, riverbed dwellers have kept a relatively low profile, enabling police and local authorities to keep a hands-off approach in many instances when public complaints were minimal, and allowing riverbed inhabitants to go about their daily routines of bathing in the river, looking for day labour, food, and clothing, and returning in the evening to the riverbed to sleep. When their visibility is raised, however (whether through discovery by passers-by or by riverbed dwellers making their presence known), police departments often subsequently engage in sweeps and arrests, usually because riverbed inhabitants are linked circumstantially to increasing property crimes such as burglaries and prowling (Sahagun 1987b).

But the increased visibility of riverbed dwellers may not necessarily result in their arrests. Depending on the context of increased visibility, the stigmatization of homelessness, both of persons and of places, may undergo radical change in a short period of time given particular circumstances. Southern Californians witnessed such a shift in perceptions concerning homeless persons in January 1995, during a period of considerable storms raging throughout Southern California. In particular, in Ventura County, many rivers flooded their banks, putting life and property in peril. For homeless persons living in river bottom encampments, such flooding not only meant loss of their homes and property, but also potential loss of life.

Homelessness in the Ventura River

A homeless encampment had long been a fixture in the Ventura River bed, where more than 200 homeless persons had constructed a set of makeshift structures, described as a 'community of plywood shacks and nylon tents below the point where the Southern Pacific train trestle straddles the river' (Alvarez 1995: A16). The scale of homelessness in Ventura County is very different from that experienced in Los Angeles. The number of homeless individuals is much smaller, and much of this population camped in the riverbed; however, business persons, police officers, and City and County officials tended to reflect similar perceptions about the nature of homelessness to those in larger metropolitan areas such as Los Angeles. Homeless persons were seen by local merchants, for example, as potentially 'damaging efforts to spruce up that [downtown] area and [attracting] tourists' (Alvarez 1996: A17). Because the homeless population tended to be confined to the riverbed, the business community, police, and City and County officials tended to have a hands-off policy, where the homeless population could be ignored because they had yet to infringe on downtown or development interests.

The situation was changed suddenly by a rash of storms, and the flooding of the homeless encampments in the Ventura River bed. In January 1995, storms caused the Ventura River to flood, placing these homeless individuals at risk of drowning. Dozens of homeless persons were rescued from the torrential flooding by rescue teams using helicopters; 'dramatic rescue scenes were played out on televisions from coast to coast' (Alvarez 1995: A16).

The near drowning and dramatic rescue of over a dozen homeless individuals spurred the City, County, and non-profit service agencies to co-ordinate assistance efforts, culminating in the establishment of a centre which matched homeless individuals with housing and social services. The river flooding and rescue was seen as a natural disaster by local authorities, and the Ventura City Council in response allocated or earmarked funds to operate the assistance centre: $13,900 for operating expenses, and $30,000 for housing and feeding 'displaced persons' for a few weeks following the flood (Alvarez 1995: A16). In addition, space was rented throughout the City to provide more than 50 shelter beds. The federal government also participated in this co-ordinated effort by promising up to 200 emergency housing vouchers for rent subsidies (with the goal of transferring individuals into permanent housing). In the first week of the centre's operation, more than 130 homeless persons were registered to participate. More than 70 homeless individuals were placed in permanent housing, while 25 were matched with temporary employment opportunities.

However, two months after the nationally televised disaster and rescue, the assistance centre was scheduled to close (on 31 March 1995). Advocates for homeless persons argued that without the dramatic scenes associated with the helicopter rescues, galvanizing support for homeless assistance would be much more difficult. As Rick Person, president of the local homeless coalition, argued, 'We don't have helicopters pulling people out of raging waters, so it's a little harder to galvanize support. But that doesn't mean it's not possible' (Alvarez 1996: A17). Service providers who had participated in the development of the assistance centre expressed concerns about the temporary nature of the centre and about the potential avenues for housing and services once the centre closed its doors. Bob Dailey, a County mental health worker, expressed his concern about the future of these homeless individuals now that the riverbed was flooded: 'My contention has always been that it's ill-advised to remove them from the river bottom when we have nothing else to offer' (Alvarez 1995: A16). Service providers were also worried that the flooding and disaster-type response, and the co-ordinated effort to develop the centre, would comprise the sole policy put forth to address Ventura's homeless population. Clyde Reynolds, executive director for Turning Point Foundation, a service agency for mentally disabled individuals, commented, 'I am concerned that once they've done this, they may feel they have done all they are legally required to do' (Alvarez 1995: A16).

Indeed, after the flood waters had receded, local policy-makers developed strategies to ensure that homeless encampments would not be allowed to re-establish themselves in their former riverbed home. New local ordinances were developed which prohibited panhandling and camping, and were implemented. Nine individuals were arrested under the new laws when they tried to move back to the riverbed. In addition, to implement the laws, the Ventura Police Department began regular patrols of the river bottom, seen by the police as a 'severe crime problem' (Alvarez 1996: A17). Lt. Carl Handy, of the Ventura Police, argued that such patrols were 'necessary to flush out homeless persons after the area was declared off-limits' (Alvarez 1996: A17). As he argued, 'It's kind of tough love stuff, forcing people to do things they don't want to do but for the right reasons. But it seems to have paid off. It's a fun area that people should be exposed to. It wasn't fun a year ago, it was dangerous' (Alvarez 1996: A17).

Fragmentation of Stigma Surrounding Homelessness

These three examples represent a structure, even if they were only implied in the three stories. The structure revolves around the issue of how stigma is constructed and expressed, given the presence and visibility of homeless persons and places, and a public which sees homelessness as undesirable. To make this structure explicit requires me to step back from these concrete examples to a more theoretical examination. Especially important in this stepping back from the concrete is how to understand the actions and motivations of local officials (whether police or elected politicians), residents, and homeless persons in the dynamic context of Southern California.

As I discussed in Chapter 2, the usual conceptualization of the rejection of homeless persons, facilities, and places centres on one of two logics, either that rejection emanates from the attitudinal structures of residents, businesses, and local political officials, or that rejection emerges from social relations of production where the NIMBY syndrome is an inevitable outcome of production processes which create negative externalities (Dear and Taylor 1982; Lake 1993).

Attitudes, Behaviour, and the NIMBY Syndrome

Instances of NIMBY are often noted in the popular media and by scholars as indicative of intolerance and prejudice in US society towards marginalized populations. Using the first conceptual framework, based in attitude formation, the local rejection of homeless persons and places, like the rejection of other marginalized groups and undesirable land uses, depends on the personality and socio-demographic traits of the rejecters, the socio-demographic characteristics of those rejected, programmatic characteristics of facilities or land uses, and, most importantly, the proximity of homeless persons and places to the individuals or groups with the power to reject. For any neighbourhood, city, or region, the landscape is conceived as a cognitive map of desirable and less desirable places with distinct configurations of possible land uses and potential users (e.g. residents and businesses). In this framework, it is assumed that if a facility associated with homelessness were to be sited in a neighbourhood, residents, businesses, and local politicians would act in ways which correspond to their socio-economic status, their perception of homeless persons and facility operators, and the social and physical structure of the neighbourhood. The resulting action by neighbourhood residents and businesses might be accepting or rejecting, and the consequent siting of facilities or homeless persons successful or a failure, depending on these neighbourhood characteristics and the behavioural intentions of those involved.

This attitudinal perspective highlights various aspects of the rejection shown towards homelessness in Skid Row, Silver Lake, and Ventura. There is little doubt that attitudes played a integral role in the rejection of homeless persons across all three examples. Residents believed that homeless persons were responsible for a perceived rise in robberies, vandalism, and general physical structure deterioration, businesses complained that homeless persons were potential obstructions to tourism and routine business practices, and police and public officials believed that homeless persons constituted health hazards and threats to public safety. Given the perception that homeless persons are very undesirable, the actions by the City of Los Angeles, the residents in Silver Lake, and the City of Ventura could be interpreted as expressing negative attitudes towards homelessness, acting to relocate homeless persons and remove homeless places as a consequence of negative attitudes and concomitant behavioural intentions. Throughout the disputes associated with the three sites, policy-makers often justified the strategies for relocating homeless persons or eradicating homeless sites in terms of the potential threats posed by homeless persons and places to residents and businesses. In all three cases, there was often little direct evidence to link the proximate homeless population with rising crime, depressed business, or deteriorating physical surroundings (and often evidence in direct contradiction to such arguments); however, the potential for such phenomena and the perception that the sole cause of these occurrences was the proximate homeless population was deemed sufficient for police sweeps, arrests, and the relocation of homeless individuals.

Using this perspective, the rejection of homeless persons and places was simply a matter of negative attitudes. Such a conclusion would indicate that the NIMBY syndrome could directly be addressed by changing people's attitudes towards homeless people and places. But this interpretation, in its straightforward search for motivations for rejection, tends to overlook the underlying and less visible sources motivating rejecting behaviour. Indeed, the attitudinal explanation has been recognized by those who reject homeless persons and places as being politically incorrect and therefore has been increasingly supplanted by other less intolerant arguments, such as the selected site is undesirable by homeless individuals themselves, or that homeless persons would be better served by a different location.

Social Relations and the NIMBY Syndrome

But the pervasiveness of NIMBY, and its application across multiple community types and locations, indicates a deeper mechanism underlying its formation and expression. The second explanation for rejection of homeless persons and places which has been promoted by scholars does not centre primarily on resident attitudes and behaviour, rather, this set of interpretations emanates from the social relations of production, where homelessness is created as an unintended but inevitable outcome of production and development (Lake 1993). Thus, locational conflict over the siting of facilities serving homeless persons and the location of spatial concentrations of homeless populations will necessarily result because production processes will always generate negative externalities such as homelessness. Class conflict arises over assigning the burden of dealing with homelessness, as community, capital, and the state struggle over the issue of who is to blame for an expanding homeless population, and the scale at which remedies should be implemented. By identifying rejection of homeless persons and places as a community problem (i.e. that the reason for the escalation in homelessness is that residents and communities are intolerant and use NIMBY tactics), the identification of homelessness as an outcome of production becomes and remains obfuscated.

Using this conceptualization, the rejection shown in Skid Row, Silver Lake, and Ventura is not merely the result of intolerant and prejudicial residents whose attitudes can be changed, but rather an inescapable drama to be played out as the negative outcomes of production affect community after community. Residents' beliefs that homeless persons steal, vandalize, and destroy property represent not just their individual belief systems, but also reflect structural systems of production and consumption integral to capitalist society. The actions taken by the City of Los Angeles, the residents in Silver Lake, and the City of Ventura could thus be interpreted as the local state acting to enable capitalist production relations (e.g. by removing homeless persons to placate the business community, and to address resident concerns about their property values and quality of life), and also working autonomously from capital and labour to fortify and legitimate its own position (e.g. by constructing local ordinances and laws to prevent homeless encampments, by using local police powers to control the rise and dispersion of homelessness, and by acting with and in contradiction to homeless advocates to provide services and aid to homeless persons).

Using this perspective, controversy over homeless persons and places is an inescapable outcome, based on the maintenance of the social relations of production and reproduction, requiring that community, capital, and the state work together or engage in conflict to eradicate homelessness (i.e. removing homeless persons from proximate locations or eliminating homeless places). While this interpretation is more satisfactory than the first as it integrates structural relations into an explanation of localized conflict over homelessness, it should be expanded to incorporate the fragmented nature of rejection. In particular, the rapid changes in local response to homeless persons indicated by the Ventura example highlight how particular circumstances may at times lead not only to lessened rejection, but to outright embracing of homeless populations.

Stigma and the NIMBY Syndrome

The perspective of production relations remains an integral component of any explanation of public response to homelessness because it highlights the pervasive nature of homelessness across time and space, and indicates why rejection of persons and places is an inescapable dimension of capitalist social relations. However, to explain the peculiarities of public response, ranging from obvious rejection to enthused acceptance, an additional conceptual dimension seems warranted. Building on the illumination developed through attitude research and explanations based in social relations, I can begin to explain and interpret the three examples in a way which recognizes the fragmentation and fragility of public response. Specifically, the herding of homeless persons and places in Skid Row, Los Angeles, to a central and less obtrusive location, the containment of vans and trailers to a one-block area in Silver Lake, and the creation of anti-camping ordinances and police sweeps in the Ventura River bed could all be construed as expression of stigma concerning homelessness.

The three stories presented in this chapter exhibit the protection of production relations (e.g. the business community believing that homeless persons would damage retail and tourist business, and residents believing that homeless persons would deflate property values), and the negative perceptions of homeless persons (e.g. persons working and shopping in downtown Los Angeles believing that homeless persons are potentially dangerous and drawn to the bars and social services associated with Skid Row). But these perceptions and actions can drastically shift, as the Ventura River flood and rescue so vividly indicate, virtually overnight. The seemingly intractable stigma associated with homelessness can be fragile, as was highlighted when homeless persons in Ventura were plucked from torrential floodwaters in dramatic rescues televised nationally. Homeless persons were not then seen primarily as the non-productive, potentially dangerous, and personally culpable individuals often stereotyped through the media and public dialogue. Rather, these riverbed inhabitants were viewed by the public and policy-makers as innocent victims of a natural disaster, seemingly no different from any other individual who might be saved from rapid floodwaters. The fragility and fragmentation of the rejection of homelessness was critically defined in these three examples by place- and time-specific circumstances, where rejection was often the widespread perception and basis for action, but enthusiastic acceptance of homeless persons could and did in fact occur.

Conclusion

This chapter, and its themes and issues, were approached as an interpretive endeavour. Hence, given three different examples of localized rejection towards homeless persons and places, I focused on interpreting the possible reasons of various actors in their varied responses. I want to emphasize, however, that I do not claim to have constructed the definitive interpretation of the three illustrative examples. Indeed, as I indicated in the discussion, there are any number of possible interpretations of the reasons for, and subsequent ways to address, the rejection of homeless persons in particular situations. The number of possible explanations in fact indicates the complexity underlying individual examples of the NIMBY syndrome directed towards homeless persons and places. And this complexity makes more clear the usefulness and applicability of attitude formation perspectives, production relations frameworks, and conceptualizations of stigma in understanding the phenomenon of public opposition.

It is clear from the description of the three cases of the NIMBY syndrome that specific events and actors were more important than others. For example, when the Community Redevelopment Agency developed multiple and increasingly costly strategies to move the Union Rescue Mission to the epicentre of Skid Row, it became clear that relocation was a primary goal of this branch of the City of Los Angeles. And, after homeless advocates vocally opposed the designation of a fenced-in drop-in centre for homeless persons in this same area, and the City and County continued to pursue this strategy (indicating that the City would change the proposed use from campground to resting area, but indicating that even more uses might be included, such as a parking lot for trailers and vans, to resolve conflicts encountered in neighbourhoods such as Silver Lake), it became apparent that while homeless advocates and activists might have input into some of the details of constructed homeless places, they would have little to say about the regional geographic scale of homeless place location.

My interpretations of the actions and motives of actors involved in these disputes were premissed upon a conceptualization of stigma that assumes a primary role for the socio-spatial relations of boundary construction and maintenance in determining the organization of marginalized landscapes. Using this account, the local politics of homelessness, which include the scale and dispersion of homelessness, the nature of homeless advocacy and activism, and the involvement of business and political agents, are central elements in the localized rejection of homeless persons and places. In this way, the spatial differentiation of the stigmatization of homelessness is intimately connected to the rise and decline of homelessness across communities, and the changing perception of quality of life and economic viability across the US. Therefore, the institutions involved in providing care for and controlling homelessness—local governments, social control agents (e.g. police), human service providers, homeless advocates and activists, and homeless persons themselves—are all part of the potential for and expression of the public rejection of homelessness.

The rejection of homeless persons and places outlined in this chapter has made more explicit how the practices of boundary construction and maintenance signify less valued persons and places. However, such an endeavour also results in the development of more valued groups and locations. That is, with the construction of marginalized groups and places is implied the development and retention of persons and places of privilege. Such privilege, however, is also fragmented and fragile, and very much dependent on the interaction between the marginalized and the so-called privileged. Such complexity is inherent when the issue of the NIMBY syndrome in communities of colour is discussed (in Chapter 7). Communities of colour have inherent within them relationships of marginalization when compared with communities defining mainstream society and encapsulate complex relationships of relative privilege, belonging, and ostracization particularly in the context of homelessness and HIV/AIDS.

7

HIV/AIDS, Homelessness, and Communities of Colour

GIVEN the rising significance of homelessness and HIV/AIDS among communities of colour, it is not surprising that in these communities, there are varied responses—some address the fall-out of these two social crises and work towards prevention, and others attempt to keep these issues from permeating further into these communities. Attempts to prevent homelessness and HIV/AIDS from further affecting communities of colour have included passive efforts (such as denial that such issues are significant) and more active practices (such as community opposition of proposed human service facilities).

Organized opposition has been an increasingly effective strategy used by communities of colour to prevent the siting or operation of these and other undesirable facilities and land uses. Researchers have generally focused their investigation of organized opposition in communities of colour on facilities and land uses associated with environmental hazards. Such research (often described within an environmental justice social movement rubric) has indicated that such activities emanate from the view that undesirable and potentially dangerous land uses are disproportionately sited in communities of colour, usually revolve around notions of equity, fairness, and justice, and are linked to broader issues such as race and class.

There are a couple of implications of this growing literature on environmental justice and environmental racism for understanding the NIMBY syndrome as it pertains to human services. First, as indicated in Chapter 5, notions such as race and gender cannot be seen as monolithic in terms of community response to controversial human service facilities. Hence, siting human service facilities in communities of colour does not emanate solely from a monolithic racism or monolithically racist institutions, nor is there a monolithic racial community response when facilities are proposed. Second, the importance of facility siting must be analysed in the context of broader issues of poverty, immigration, employment, crime, housing, and other daily needs that many communities of colour cope with on a daily basis. It is clear that human service facility siting does not constitute an overriding priority in many communities of colour, when their members

must deal with more pressing issues. However, vehement community
response remains an increasingly important option for such communities
perceiving a significant threat posed by proposed or operating human
service facilities. It is argued in this chapter that the response by communities of colour to facilities associated with homelessness and HIV/AIDS is
tempered not only by wider societal views about the undesirability of
homelessness and HIV/AIDS, but also by the specific ways in which
different communities of colour define their 'imagined community' boundaries (following Anderson 1983). The constitution of the social and spatial
boundaries of such imagined communities illustrate not only the mechanisms by which stigma is developed and expressed, but also the ways in
which such boundaries are permeable.

This argument is explored through an analysis of fifteen in-depth interviews with Latino/Latina[1] and Vietnamese informal opinion leaders in
Orange County, California. Other racial groups, particularly African
Americans, comprise very small proportions of the County's population
and consequently are not included in this discussion.[2] The interviews
emphasized these community leaders' views about the nature of HIV/
AIDS and homelessness in their communities, the actual and potential
community response to human service facilities, and their personal experience with these two conditions. HIV/AIDS and homelessness are perceived
somewhat differently by the two groups of leaders, but, in general, both
groups see these two crises primarily as problems for other racial groups.
Despite efforts by human service providers and advocates particularly
during the early 1990s, many community leaders appear unaware of the
significance of HIV/AIDS and/or homelessness, or remain unwilling to
confront these two issues.

HIV/AIDS, Homelessness, and Race in Orange County, California

Orange County, California, located south of Los Angeles, is not widely
known as a place of cultural diversity. Southern Californians often think of
life south of 'the Orange Curtain' as consisting of theme parks (e.g. Disneyland and Knott's Berry Farm), high-tech economic growth, master-
planned communities, suburban affluence, and conservative politics (e.g.
Scott 1986, 1993). But life in Orange County has not been static. One
reflection of its dynamic character has been the rapid diversification in its
population over the past decade. For the County as a whole, the 1990 US
Census indicated that its thirty-one cities and unincorporated areas had a
population of 2.4 million, having increased approximately 25 per cent over
the past decade. While the County's population remained dominated by
non-Hispanic Whites (about 65 per cent of the County's population in
1990), the growth in the various minority populations constituted a

primary source of overall County increases since 1980.[3] The proportion
of Asian and Pacific Islanders grew, for example, from 4.8 per cent of the
total population in 1980 to 10.3 per cent in 1990, representing a 166 per
cent change over the decade. The Vietnamese community, in particular,
experienced rapid growth during the 1980s, more than doubling in size,
and currently representing about 3 per cent of the County's total population. The Vietnamese population tends to be concentrated in the cities of
Garden Grove and Westminster (in the north-west portion of the
County).

While the Hispanic population in Orange County did not grow as
rapidly as many of the Asian Pacific groups, their proportion of the
population increased from 15 per cent in 1980 to 23 per cent in 1990.
Hispanics in Orange County (comprised mostly of individuals of Mexican
descent, and usually self-identifying as Latinos and Latinas) now total
approximately 565,000 individuals. The County's largest city, Santa Ana,
with a population of approximately 294,000 was about 65 per cent Latino
in 1990. The Latino population tends to be concentrated in the northern
County cities of Santa Ana and Anaheim. In general, there is relatively
little interaction between the Latino and Vietnamese populations in
Orange County, although there are growing professional affiliations
through business associations and human service providers.

The significant changes in the demographic portrait of Orange County
have implications for its racialization of work and residence. A central
element in the economic development of the County, particularly between
the 1950s and 1980s, was the growth in aircraft and defence-related
industries (Scott 1986, 1993). The resulting high-technology industrial
complex created a polarized labour market, where highly skilled engineers,
technicians, and management tended to be white, while the blue-collar
labour force (involved in agriculture, construction, and benchwork)
provided employment for the growing Latino and Asian immigrant
population (Rogers and Uto 1987). Recently, the cutbacks on defence-
related spending (resulting in the closing of military bases and the down-
sizing of defence-related manufacturing and production), the fiscal woes of
the County (suffering recently from the largest municipal bankruptcy in
US history), and the growing political focus on immigration have meant
that low-income households, particularly recent immigrants, have been
subject to a shrinking labour market, pressures to reduce or eliminate
health and human services, and an increasingly vitriolic public response
to their presence.

While few understand Orange County as a region of rapid demographic
change, even fewer (both within and outside the County) perceive this
municipality as having to deal with issues of poverty and homelessness.
As in other metropolitan areas, however, particularly during the late 1980s,
homelessness became an increasingly significant issue in the County.

Homeless advocates in the County estimate that there are between 12,000 and 15,000 homeless persons throughout the County on any given night (Dolan 1995). Perhaps because of the lack of recognition of homelessness as a significant issue, there remains relatively little policy intervention regarding emergency shelter, affordable housing, or other issues specifically associated with homelessness. Across the County, for example, there are only about 1,000 shelter beds available (Dolan 1995). There has not been a co-ordinated County effort to address homelessness, rather programmes and facilities tend to operate in an *ad hoc* way, with many programmes extremely vulnerable to the changing nature of federal, state, and private charitable funding sources. Facilities providing services for homeless persons tend to be concentrated in the northern, more urbanized areas of the County; however, there are food banks, drop-in mental health care clinics, and other services sporadically located throughout Orange County.

Research on homelessness in Orange County has indicated that it is both similar and distinct from homelessness occurring in urban, metropolitan, 'inner-city' areas, such as the Skid Row areas of cities (Crane and Takahashi 1997; Cunningham *et al.* 1993; Mishra 1994). Similarly to studies of inner-city homeless populations, samples drawn from the Orange County homeless population are largely male and unmarried, with large proportions newly homeless (i.e. less than one year). However, homelessness in the County also tends to reflect different demographic patterns from inner-city populations, with large proportions White and Latino/Hispanic and smaller proportions consisting of African Americans (although African Americans still tend to be overrepresented in the homeless population when compared to the US Census describing the overall County demographic portrait). A large proportion of these homeless persons have lived in the County for long periods of time, many living in the area for five years or longer (Crane and Takahashi 1997).

HIV/AIDS has also become increasingly significant in Orange County since the late 1980s.[4] Between 1987 and 1993, over 1,500 deaths in Orange County were attributed to HIV/AIDS, with a steadily increasing trend in the number of deaths.[5] The Orange County Health Care Agency estimated that between 6,400 and 8,600 persons were living with HIV/AIDS as of January 1994. People living with HIV/AIDS in Orange County tend to be overwhelmingly male (approximately 90 per cent) and primarily White. Women constituted approximately 24 per cent of the population testing HIV-positive in 1994. Although the population of non-Whites living with HIV/AIDS remains relatively small, the proportion of Latinos/Latinas, African Americans, and Asians living with HIV/AIDS has increased, particularly over the past five years. The proportion of AIDS cases by Latinos/Latinas rose from 16 per cent in 1989 to 26 per cent in 1994, the proportion of African Americans increased from 3 per cent in 1989 to 5 per cent in 1994, and the proportion of Asians rose from 1 per cent in 1989 to 3 per cent in 1994. These 1994 statistics indicate that, in Orange County, Latinos/Latinas and African Americans were disproportionately represented among reported AIDS cases, and Asians were underrepresented, mirroring national statistics collected through the Centers for Disease Control and Prevention (CDC 1996).

Public perception of HIV/AIDS in the County as a whole is one of concern, tempered by its lesser priority in relation to gangs, violence, and substance abuse. A recent attitude survey of Orange County residents about AIDS education and prevention indicated that most respondents see AIDS as less problematic than crime, substance abuse, and health care for low-income persons, that most believe that AIDS education should be incorporated into high schools, and that few (about 10 per cent) would participate in protests against AIDS education and prevention programmes in the public school system (Baldassare and Katz 1995). Such moderate political response tends not to be the norm for a county dominated by conservative politics and which served as the birthplace for Proposition 187 (a recently passed California State initiative which would deny education and health benefits to undocumented immigrants and their families).

The institutional response to HIV/AIDS has tended to be relatively progressive even given the conservative nature of the County's politics. The Health Care Agency of the County of Orange has, since the early 1990s, used applied anthropology to develop culturally specific prevention and education programmes for female sex workers and undocumented Spanish-speaking males (Carrier and Magaña 1992). In addition, as a consequence of changing federal procedures for determining funding priorities and disseminating funds, the County has co-ordinated an HIV Prevention Planning process incorporating multiple perspectives: persons living with HIV/AIDS; persons of colour; gay, lesbian, bisexual, and transgender persons; service agencies; academics and professional consultants; and multiple County agencies (Orange County HIV Prevention Planning Committee 1995).[6] Human services funded through the County associated with HIV/AIDS provide HIV testing, referral, social support, or case management services. These services tend to be concentrated in Santa Ana (the County seat) or Laguna Beach (a gay/lesbian enclave), but are also located in other cities.

Exploring Community Response

To investigate the construction of stigma towards facilities associated with HIV/AIDS and homelessness in communities of colour in Orange County, fifteen detailed interviews were conducted with Latino and Vietnamese informal opinion leaders in Orange County. Informal opinion leaders are individuals whose advice and attitudes have the potential to affect behaviour significantly, without the formal powers of institutionalized government or widely acknowledged political positions. The importance of informal leaders on the attitudes of residents may be paramount in terms of influencing the construction of stigma and subsequent community response towards facilities associated with HIV/AIDS and homelessness.[7]

Opinion leaders were identified using local newspaper articles and interviews with researchers and community leaders. A snowball method was used to identify Latino and Vietnamese persons holding significant public influence. Local newspaper articles and researchers were used as initial contacts to identify seven individuals. These seven individuals were interviewed about the nature of HIV/AIDS and homelessness in the Latino and Vietnamese communities in Orange County, and were asked to provide names of individuals who were influential in these two communities of colour. From these seven individuals, twenty Latino and Vietnamese informal opinion leaders were identified and contacted for interviews. Fifteen agreed to be interviewed. The group interviewed was limited to those speaking English because the author was unable to converse in Spanish or Vietnamese. This constraint probably biased the results of this analysis towards more acculturated individuals.[8]

The persons interviewed constituted members of professional networks linked with researchers, community leaders, and the local media. They did not comprise a random sample of individuals in these communities, and thus should not be considered a representative sample of Vietnamese or Latino attitudes in Orange County. The usefulness of this group of individuals lies in their informal influence within the Vietnamese and Latino populations. Many of these individuals are highly respected religious or civic leaders. At the very least, they are very well connected to various subgroups within two communities of colour in Orange County.

The interviews took place at the person's primary place of work (e.g. office, church, or facility) and were conducted privately with each respondent. The length of the interviews ranged from thirty minutes to two hours, and were conducted between 14 September and 5 October 1994. Each of the interviews was taped and later carefully transcribed, paying attention to transcribing the conversations accurately. The interviews were semi-structured (based on a set of questions determined prior to the series of interviews), and informal. The set of questions was asked of all those interviewed; however, each of the individuals was encouraged to talk about related issues. The questions used in the in-depth interviews were based on research findings in the literature on minority attitudes towards HIV/AIDS (e.g. Airhihenbuwa *et al.* 1992; Flaskerud and Uman 1993; Marin 1993), and geographic exploration of community acceptance of controversial human service facilities (e.g. Dear *et al.* 1997; Smith and Hanham 1981). Two distinct but interrelated facets of the stigmatization of HIV/AIDS and homelessness formed the core of the semi-structured questions: persons living with HIV/AIDS and homeless individuals (e.g. characterization and community acceptability); and facilities associated with HIV/AIDS or homelessness. The facilities were not described in detail, but the Vietnamese and Latino/Latina respondents generally referred to hospices or group homes when they spoke about community response to human services associated with HIV/AIDS and homeless shelters when they spoke about homelessness. If respondents queried the nature of facilities, they were provided with a brief description of general facility types (e.g. drop-in facilities with daily user turnover, residential facilities providing housing and services to persons living with HIV/AIDS or homeless individuals). Only a few respondents queried the nature of the human service facilities associated with HIV/AIDS or homelessness.

Respondents were selected for their leadership roles in the community as either religious leaders (7) or health and social service leaders (5). One person was from the private sector. No attempt was made to determine views or personal experience of HIV/AIDS or homelessness. Attempts were made to interview an equal number of women and men; however, because of the snowball method of selecting persons to be interviewed, most of the respondents were male. Ten of the respondents were Latino/Latina and five were Vietnamese. None of the respondents identified themself as gay, lesbian, or bisexual. Only one person was a recent immigrant (i.e. fewer than 5 years in the US). Most of the respondents were in their 40s, but ages ranged from the 30s to the 60s. None of the respondents are identified by name to preserve the confidentiality of the views expressed.

Imagined Communities, Stigmatization, and HIV/AIDS

The imagined community of HIV/AIDS as consisting of gay men was central in the definition of safety and deviance described by the Vietnamese and Latino/Latina respondents. Indeed, while homosexuality was not included in the set of questions posed to the respondents, all the respondents referred to homosexuality when questioned about the nature of HIV/AIDS in their communities. Many of the conversations turned to the unacceptability of homosexuality in the Vietnamese or Latino communities in the midst of the discussion concerning the existence of HIV/AIDS in these two communities.

they [community members] blame it [HIV/AIDS], and seem to continue to attach it to lifestyle, to the fact that it's a gay disease. (Latina female, 40s, assistant to County supervisor)

Homosexuality for the Vietnamese and Latino/Latina respondents was largely defined in terms of abnormality and deviance. Homosexuality for some of those interviewed signified any deviance from social norms. Deviance did not have to be extreme to warrant the label of homosexuality; marrying at an older age than traditional conventions for example could trigger the label 'homosexual'. For most of the respondents, however, homosexuality was understood as a marked difference from acceptable norms, leading to a spoilt identity, and associated with shame, disgrace, or infamy. Respondents indicated the importance of the perceived essence of the difference from socially established norms in their understanding of homosexuality and HIV/AIDS.

You start, you learn, you know, it was brought up in my, my parents that, you know, right, right from high school, you need to work. And that's, that's, you need to provide. You are a provider. The important thing, if you don't work and go to school, then there's something wrong with you. You're gay. You're, you're abnormal. Something is wrong with you. You're not, you know, or you don't get married or bear children. There is something wrong with you. I didn't get married until twenty, twenty, when I was twenty-three and then my, they were questioning, you know like, well, are you able to get married, or are you, you know. (Latino male, 30s, executive director, non-profit organization, non-human service)

Homosexuality is a very strange thing. Sex, female sex and men sex are too different, and they are made there to another sex. They are not to the same sex. They [young Vietnamese persons] think that is very, very funny. (Vietnamese male, 40s, Catholic priest)

This construction of HIV/AIDS as homosexuality operated in two ways. HIV/AIDS as homosexuality allowed the denial of the existence of HIV/AIDS in the Vietnamese or Latino community by the respondents, and it necessitated and reinforced the invisibility of persons living with HIV/AIDS in these two communities of colour. Denial was a common response to the question of the existence of HIV/AIDS in the respondents' communities, especially by the Vietnamese respondents. Much of the denial of the significance of HIV/AIDS, however, was tempered by an acknowledgement of its growing importance or its invisibility in these two communities.

No AIDS problem. I live here for 14 years. I never heard AIDS problem. (Vietnamese male, 60s, Buddhist abbot)

It is not seen. They [persons living with HIV/AIDS] are not brought out in the street. They are hidden. (Vietnamese male, 40s, executive director, housing/human services)

AIDS is still a kind of problem under the blanket. Those of us who are in leadership positions are probably more aware than the general population. (Latino male, 40s, Catholic priest)

The invisibility of HIV/AIDS interacts with the culturally specific definitions of homosexuality and homosexual behaviour. Several of the Latino respondents indicated that men who have sex with men do not necessarily self-identify themselves as homosexual, adding to the complexity of the issue of invisibility with respect to HIV/AIDS. The lack of self-identification as homosexual may emanate from denial and fear of retribution, or from culturally specific definitions of the constitution of homosexuality and its connections to HIV/AIDS (Magaña and Carrier 1991). That is, the construction of HIV/AIDS as homosexuality in the Latino community meant that the risk of transmission of HIV/AIDS only existed for men labelled 'homosexuals', and not necessarily for men who engaged in same-sex behaviour.

The understanding of the whole sexual activity is that only 'maricon' get AIDS. Even if I have a homosexual contact I assume the male [dominant] position and therefore I will not get AIDS because I am still a man. That whole idea is still very much out there which is related to the whole weakness thing. You know, he's not a real man and this is why these kinds of things befall him. (Latino male, 40s, Catholic priest)

With the widespread belief that HIV/AIDS is confined to the homosexual population, individuals engaging in heterosexual behaviour perceive themselves as not being at risk, and may then behave according to this belief. Such widespread understanding of HIV/AIDS tends to interact with moral standards for behaviour held by respected social institutions, resulting in unwanted and increasingly dangerous outcomes.

The culture states that men may actually be, you know, out playing around, but the Church doesn't necessarily condone that they protect themselves. You know, so you've got this conflict of maybe somebody going out, but not necessarily taking precautions and picking up a disease. (Latino male, 40s, physician)

The social construction of HIV/AIDS as homosexuality and the subsequent invisibility of HIV/AIDS within these two communities of colour created struggles for preferred group association. This decision for group association necessitated a choice between a continued alliance with a person identified as having HIV/AIDS and a continued connection with the Vietnamese or Latino community. At times, the continued connection with the Vietnamese or Latino community required an ostracization of persons living with HIV/AIDS. If a person was not identified as living with or dying of HIV/AIDS, then continued linkage with the Vietnamese or Latino community remained a possibility, either by non-disclosure or by the explanation of illness or death as the result of some other condition, such as cancer or substance abuse (also Aranda-Naranjo 1993). The rising significance of HIV/AIDS and knowledge about transmission and risk groups means that conditions such as cancer and substance use may be seen as more acceptable than HIV/AIDS (Takahashi and Dear 1997). The use of any of these coping mechanisms, however, was painful and distressing, as families tried to cope with the difficult choice between continued association with a person with HIV/AIDS or the maintenance of ties to their communities.

Even within the family, just thinking of my cousin [who died of AIDS], it's just, move away. He had to move very far away because the family, you can change, just change, and you know, he couldn't, so he was ostracized. He was loved, he was grieved, because the family still longed for him, but they couldn't bear to have, and again you don't even know if his own mother in the back of her mind is accepting, well, or in the back of the mind accepts the fact that he died of AIDS. You know, to hear her describe his death, because she was the only one that went back, she went back to be with him at his death. She describes it in a way, and we knew from his friends what was going on, just, as though it was not AIDS, just something. (Latina female, 40s, assistant to County supervisor)

I am not sure if that is still true, but I remember one time a woman was coming to me about her husband [who] had died of AIDS and she pleaded with the doctor to record it as he died as an instance of drugs. That he was a drug addict. (Latino male, 40s, Catholic priest)

The religious and cultural norms which compel the ostracism of persons with HIV/AIDS, however, may also work to promote sympathy and caring when HIV/AIDS is understood as 'illness' rather than 'homosexuality'. Religious and cultural norms dictate the exclusion of unacceptable behaviour; however, in specific situations, the caring facet of religious doctrine can override more exclusionary dictates with respect to persons living with HIV/AIDS.

Well they [Vietnamese people] feel sympathy towards people with AIDS. They care about them. It is not just because of culture, religion taught us we must love each other and have sympathy, especially Catholic or Protestant. . . . There is no reason to look down on AIDS patients or insult people who are sick. There may be some conservatives who think differently who think it is because of extra sexual relationship. (Vietnamese male, 30s, Catholic priest)

I think that people are very compassionate to their relatives. For example, the family that I met, even though this, ah, young man who was 29 years old, even though he had been married in the past, and had been using drugs, and stealing from them, and doing so many things, they still took him back. They were taking care of him. (Latino male, 40s, Catholic priest)

Caring for family members living with HIV/AIDS in these two communities is also facilitated by the denial of the condition. The person living with HIV/AIDS can remain a part of the family and larger community (because her/his condition is due to some relatively acceptable source), and the family can continue to provide emotional and material support without fear of retribution from their community, wider society, or the church. Denial also eliminates the need for confrontation and conflict concerning the understanding of HIV/AIDS as homosexuality.

HIV, AIDS, that's you know, I've dealt with maybe, personal, that whole area is clouded, or covered over, or not accepted. I think there's a lot of denial. I'm dealing right now with two young men that are dying of AIDS, and neither of the families has admitted [that the men have AIDS], or they claim it's another illness. They guys themselves have told me they have AIDS, . . . but one sister said 'oh, he has herpes' or 'he's very sick with his nerves'. The mother, she doesn't know. (Latino male, 40s, Catholic priest)

The imagined community of HIV/AIDS as consisting of gay men has brought with it the cost of ostracization. The imagined community defined by the Vietnamese and Latino/Latina respondents as 'their' communities, on the other hand, signified spaces of safety and protection. If members of the Vietnamese or Latino communities followed traditional cultural practices and did not engage in Western or American behaviour, then, respondents argued, these community members would be safe from the dangers associated with HIV/AIDS. This perspective was especially prevalent among the Vietnamese respondents, who were for the most part members of the clergy. They argued that tradition, culture, and community were adequate and effective shields against the perceived invasion of HIV/ AIDS emanating from the less structured Westernized society. The Vietnamese respondents with religious affiliations viewed the invasion of HIV/AIDS not only as a health issue, but also as a potentially destructive element threatening the cohesion and stability of their community.

Now, I teach the Vietnamese Catholics here about religion; they are very, very religious. And so, I don't think they have AIDS. Because, intense to religion for the country to homosexuality is strange thing to our culture. They feel that when they hear about AIDS, they feel that this is a very funny problem. . . . Because so far in the Vietnamese Catholic community there's no cases of AIDS and so they don't feel anything because there are no cases and if there's any cases then it's very secret. It's not a problem for them. AIDS a problem for other people but none for the Vietnamese Catholic. (Vietnamese male, 40s, Catholic priest)

They [Vietnamese people] are aware of the danger [of HIV/AIDS]. I think it is because it reflects a clear distinction between the Asian values, moral values, and the Western moral values. It involves the free sex, the careless sex that is not commonly heard [of] in Vietnam. (Vietnamese male, 50s, Lutheran minister)

Acculturation and immersion in Western culture and values played a complex role in the social construction of HIV/AIDS in these two communities of colour. Among many of the Vietnamese respondents, there was the perception that HIV/AIDS was linked to younger, more acculturated Vietnamese who believed less fervently in traditional values and were more likely to engage in risky behaviours (e.g. especially with respect to sexual practices). Persons with fewer ties to Western culture were perceived to be far less likely to become influenced by these less structured norms, and, therefore, were argued to be far less at risk of contracting HIV/AIDS.

I think those who have been here longer usually become more and more careless about sex matters, but those who are new refugees, well, because they have no way to mingle quickly with society. So they are usually confined to their families, to their cultures, to their traditions, and that is why they can avoid this disease. Not because of their will, but because they are taught and they are raised to be within the family. If they are run away, they stay with their peers, Vietnamese peers. Usually, the Vietnamese children don't mingle with other children of other ethnic groups. So there is culture dividing that somehow helps them stay away from HIV and AIDS. (Vietnamese male, 50s, Lutheran minister)

The imagined community offering protection and safety to community members was conceptualized as culture and traditions drawn from Vietnam and Mexico, Latin/South America. These ties were made complex, however, by the changing context of HIV/AIDS in these nations. In the case of the Vietnamese respondents, the construction of their imagined community expressed through Vietnamese culture and tradition was being threatened by the increasing incidence of HIV/AIDS in Southeast Asia. While Vietnamese living in the US might be protected from HIV/AIDS through the traditions and cultural practices espoused by moral leaders, such practices have not proven to be effective in Vietnam. Consequently, although HIV/AIDS as homosexuality and its concomitant insignificance for Vietnamese persons continued to dominate the social construction of HIV/AIDS among the Vietnamese respondents, there was some doubt expressed about this definition. But even with this growing concern about HIV/AIDS in South-east Asia, Vietnamese respondents expressed confidence that living in the US provided protection because of institutional and political barriers to immigration of persons living with HIV/AIDS. Thus, the Vietnamese respondents indicated that their localized community may be protected from the invasion of HIV/AIDS even though HIV/AIDS was becoming significant in South-east Asia.

They [Vietnamese people] know, because it [HIV/AIDS] is not their problem. They know about that. They know that is a problem, but they are not concerned. They tell me now in Vietnam, in my country, there are cases of AIDS now. . . . I think AIDS people can't come to America because they have screening in Vietnamese by the America office over there. (Vietnamese male, 40s, Catholic priest)

The understanding of HIV/AIDS as an issue for individuals outside the Vietnamese and Latino communities was intricately linked to the respondents' perception of the acceptability of human service facilities associated with HIV/AIDS. There were dual dimensions affecting the acceptability of such facilities. One dimension was the stigmatization of HIV/AIDS and persons living with HIV/AIDS emanating from the understanding of HIV/AIDS as homosexuality. However, caring and nurturing constituted a countervailing dimension dictated by religious and traditional norms when HIV/AIDS was understood as illness. These two dimensions meant that the Vietnamese and Latino/Latina respondents expressed both accepting and rejecting views when queried about their communities' response towards facilities serving persons living with HIV/AIDS.

If they know that facility[9] has AIDS people, I don't think that they would come there to live in that facility. (Vietnamese male, 40s, Catholic priest)

I guess the impression for the most part is that most people who have AIDS are homosexuals and then that would lead to more homosexuals in your backyard. (Latino male, 50s, business owner)

No problem at all. Very, very welcome to have a centre [for persons living with HIV/AIDS]. But we should have enough staff who have understanding and provide appropriate services for these patients. (Vietnamese male, 30s, Catholic priest)

Acceptability for facilities associated with HIV/AIDS was strongly influenced by the role that such facilities might play in neighbourhoods. Increasingly, communities of colour have become involved in locational politics because of the view that undesirable facilities are disproportionately sited in low-income neighbourhoods predominantly populated by persons of colour (Bullard 1990; Pulido 1996a). Acceptance for human service facilities associated with HIV/AIDS therefore rests in part on the acknowledgement of HIV/AIDS as an issue significant for particular communities. Given that most of the opinion leaders interviewed did not recognize HIV/AIDS as a problem for their communities, there was not a recognized need for such facilities in close proximity. Thus, the siting of any facility for persons with HIV/AIDS might be seen as an attempt to site a controversial and undesirable facility in contradiction to the needs and desires of community members. Many respondents, while indicating that most members of their communities did not recognize the significance of HIV/AIDS, did suggest that there was a growing awareness of HIV/AIDS within their communities.

I don't think it would be acceptable if it [a facility associated with HIV/AIDS] was marketed that we are going to put it in your community because that is where the land is cheap. So one it has to be coming into the community that is presented as something valuable, not just to the community but the community at large. So that would be one thing. The other thing would be that we are here to help you, it's a problem in your community, you all know it, and this will be a place where we can actually help you with this situation as problems develop, rather than this is an AIDS centre and this is going to be in your community. I don't think that would go over as well. So a lot of it is how it is marketed, it has to be marketed for the community, the community has AIDS, everybody knows it, they don't talk about it, so you're not going to make a big deal about it, yet you are going to address the needs of the community and you are going to keep people out of the hospital. (Latino male, 40s, physician)

Associated with the perceived need for human service facilities associated with HIV/AIDS was the issue of power and control over HIV/AIDS and facility siting within both communities. For some of the respondents, the siting of a controversial human service facility in close proximity to residences and businesses, especially one for which their communities had little or no need, constituted an overt expression of marginalization and oppression. These respondents argued that the suspicion of marginalization would intensify if the human service facility were operated and used by outsiders and strangers to their community.

The idea of an outside agency coming in and imposing that, no matter what kind of program you are involved in, they are going to be suspicious because we don't know if it's the city, or it's this or it's that, we don't know these people. (Latino male, 40s, Catholic priest)

Control was also relevant at the individual level. Human service facilities might prove undesirable or irrelevant for persons who do not believe they have any measurable control over HIV/AIDS transmission. The power relations in households therefore constitutes a critical factor in access to human services and individual use of knowledge and information concerning HIV/AIDS. For Latina women, in particular, gender and sexuality play central roles in their desire for information and services, and, perhaps, in their acceptance of facilities associated with HIV/AIDS. Even when Latina women are provided with information about preventing HIV/AIDS transmission (e.g. through condom use), they might not be able to act on this information given the implications of requesting such a strategy (e.g. that a husband is unfaithful, has engaged in same-sex behaviour, or is an injecting drug user). Thus, facilities associated with HIV/AIDS may not only represent threats to the imagined community of Latinos and Latinas, in general, they might also represent threats to social relationships within individual households.

The women don't want to hear about it [HIV/AIDS], plus the fact that even if they did hear about it, they couldn't do much about it, so you're stuck. (Latina female, 40s, assistant to County supervisor)

The Vietnamese and Latino/Latina respondents believed that leadership endorsement of facilities associated with HIV/AIDS would be crucial to community acceptance. All of the Vietnamese and Latino/Latina respondents indicated the importance of the church or institutionalized religion in the decisions made by persons in their communities, especially with respect to controversial human service facilities associated with HIV/AIDS. Religious institutions are diverse in these two communities, and include Catholics and Protestants for both Latino/Latina and Vietnamese respondents and Buddhists for Vietnamese respondents. These two communities are not monolithic, so these religious institutions do not necessarily control the beliefs of the entire community. However, without the endorsement of such institutions, there is little doubt of the central role that the social construction of HIV/AIDS as homosexuality would play in structuring community response to facilities associated with HIV/AIDS.

I still think that the Latino community is so tied in to the church, even the Protestant church. But a church is an important part of your, of who you are, that would have to be, the church needs to take more leadership in that. And that's the only way, to me, that people will accept [facilities for persons with AIDS]. I mean the church becomes like a mom and dad, and if mom and dad say it's O.K., then it's O.K., and so the church has to be there to say it's O.K. (Latina female, 40s, assistant to County supervisor)

Homelessness and Community Boundaries

The imagined community of closely knit, immigrant kinship and social network ties also played a central role in the Vietnamese and Latino/Latina conceptualization of homelessness. There remains a strong perception among leaders in both communities that the strength of family and kinship ties is a guarantee of housing and shelter options for those without the financial resources to obtain their own housing. The leaders interviewed all believed that homelessness was not an important issue for the Latino/Latina and Vietnamese communities in Orange County. Instead, homelessness was argued to be prevalent among those outside their communities.

Civic Center [in downtown Santa Ana, the county seat] is a great place to find any kind of homelessness you want. Just go out there—you don't see Latinos out there. (Latina female, 40s, assistant to County supervisor)

Indeed, the community leaders interviewed stressed that the community's association with homelessness lay primarily in community members providing help, services, or food to homeless persons, and these homeless persons were characterized as from outside their communities. Because homelessness was not part of the imagined community of the Vietnamese and Latino/Latina community leaders, their definition of homelessness as outside the confines of their communities tended to be reinforced by such interaction.

Now, we have a group who takes care of the homeless in Santa Ana. They come here from here to Santa Ana to give food and to serve meals on every Friday. These people are Vietnamese, but the homeless people they feed are not Vietnamese. (Vietnamese male, 30s, Catholic priest)

The community leaders interviewed provided multiple explanations for their community members not being or becoming homeless. Many of these explanations revolved around the heightened productivity they associated with recent (im)migrants and their perception of the strength and support provided through familial and kinship social ties. Hence, those from within the family or community should be able to access the emotional and material support (including housing resources) available through the extensive social networks available in the communities, and, moreover, the ethic of productivity meant that the use of such support was a temporary measure for both the receivers and providers of support.

You're not going to [find Latinos on Skid Row]. You're not going to because it's not an acceptable thing. The work ethic is there among the first generation, new arrivals. . . . So you are not going to find a lot of people out in the street, because they are going to be trying to get a job. . . . I think it is expected within our culture that you are going to be contributing and contributing through employment and through work and you are not really going to be a drain to anyone. . . . That's just unwritten rules. (Latino male, 40s, physician)

Individuals who are unable or unwilling to play by these unwritten rules are also deemed to be distinct from their communities' members, obtaining varying labels of stigma (e.g. 'divorced' or 'runaways'). In not playing by the rules, Vietnamese and Latinos/Latinas living in Orange County are faced with obstacles and boundaries which expel them from the status of community member to outsider. And though this status of outsider is not as stark or distant as the status of outsider attached to individuals not visibly Vietnamese or Latino/Latina, it still carries a stigma which prevents individuals from accessing the social support, as argued by the community leaders, which prevents homelessness in these two communities. Especially for Vietnamese and Latino/Latina persons who are homeless, the fact that they are not accessing the ubiquitous social support available throughout the communities means that they are personally culpable for their becoming, or, more importantly, their remaining homeless.

There may be some homeless cases because they came here alone, or they are divorced or they are teenagers who come alone or they came here with family and don't deal with rules that the family set up for them, so they ran away. In these cases, they might become homeless. (Vietnamese male, 30s, Catholic priest)

Conclusion

Both HIV/AIDS and homelessness are largely defined as outside community boundaries. The construction of HIV/AIDS as an imagined community of gay men has served to define powerful boundaries between safe or normal communities and communities of deviance and marginality. For individuals having to cope with family members or friends living with HIV/AIDS, the consequences of the construction of these boundaries are often very painful. Disclosure of HIV/AIDS requires that family and friends choose their preferred group association, either between the linkages to deviance and abnormality that HIV/AIDS represents or the safety and normality represented by the normal community (e.g. association with the Latino/Latina or Vietnamese communities and not with the person living with HIV/AIDS). The boundaries constructed between those homeless and those not homeless are similarly defined by normality and community practices. Prevailing understanding of the imagined communities promoted by Vietnamese and Latino/Latina community leaders is that productivity and social network ties guarantee that community members will not become homeless, or, if they do, that homelessness will be a temporary condition.

But while the Vietnamese and Latino/Latina respondents indicated the centrality of the social construction of HIV/AIDS and homelessness as outside, they also described the caring and nurturing that takes place within these two communities of colour as they struggle to maintain normality. Thus, although the construction of boundaries between normality and deviance requires the ostracism of persons identified as living with HIV/AIDS or homeless, individuals have created coping mechanisms (e.g. denial and differing diagnosis, and the social construction of HIV/AIDS as illness) to maintain social support ties while retaining their membership in the safe and normal communities of colour with which they have identified themselves. These coping strategies are important indicators of ways in which persons of colour have overcome the obstacles of community pressures for normality to continue to provide emotional and material support. These strategies indicate that, even with highly structured norms, families have found ways to maintain socially acceptable ties. This maintenance of socially acceptable ties, however, is painful and distressing, mainly because of the denial that families and individuals must undergo to maintain the invisibility of conditions such as HIV/AIDS and homelessness.

In the particular case of HIV/AIDS, while it remains highly stigmatized, facilities associated with HIV/AIDS may not necessarily experience community rejection. There are countervailing dimensions of stigma consisting of the rejection associated with the construction of HIV/AIDS as homosexuality and the caring and acceptance associated with the construction of HIV/AIDS as illness. Religious institutions may play the critical role in determining the importance of these differing social constructions of HIV/AIDS. According to the respondents, representatives of churches and temples remain highly influential leaders in defining normality and acceptance, and, therefore, they remain an important source of response towards persons living with HIV/AIDS and facilities associated with HIV/AIDS. The top-down influence of religious institutions on the social construction of HIV/AIDS appears to remain significant in large measure even for younger and more acculturated individuals.

Although respondents believed that the endorsement of facilities associated with HIV/AIDS or homelessness by religious leaders would be crucial to community acceptance, they also indicated that local control over the operation of facilities and the composition of users comprised an important facet leading to community acceptance. The need by community members for a facility associated with HIV/AIDS or homelessness would have to be established for community acceptance to occur. Establishing the need for such facilities is problematic given the widespread view by the Vietnamese and Latino/Latina respondents that HIV/AIDS and homelessness are not significant issues in their communities. However, establishing that a local need exists is vital to addressing the concern that operators and users of such facilities would be outsiders or strangers. Outsiders and strangers not only represent intrusions in the social relationships within the respondents' communities; the operation and use of facilities by outsiders also implies the further marginalization of these two communities, and a consequent lessening of control over daily life.

Part IV

Homelessness, HIV/AIDS, and Public Policy

8

Relocating Homeless Persons

The Anti-Camping Ordinance in Santa Ana, California

> If an officer sees someone standing or sleeping on property to which that person has no legal right, this in itself is sufficient evidence for both arrest and conviction.
>
> Robert Wennerholm, Police Law Institute (1994)

ANY understanding of the relationships between the NIMBY syndrome and homelessness must consider the role and status of municipal governmental bodies in the management and control of homelessness.[1] It is this level of governance that is responsible for the administration and interpretation of cold weather shelter policies, for policing strategies and enforcement of municipal codes and ordinances, and for the control over siting of homeless shelters and human service facilities. Further, it is the decisions of municipal governments which have had profound impacts on the health and well-being of homeless persons, the success and failure of organized community resistance to homeless shelters, group homes, and affordable housing, and the capacity of human service providers to deliver programmes and services to needy populations, just to name three related facets.

Municipal governments have become increasingly responsible for their homeless populations. Welfare state reorganization and continued efforts at new federalism have made states and localities more responsible for maintaining a safety net for those affected by a lack of affordable housing, deinstitutionalization of mental health care, reductions and changing eligibility for federal and state welfare benefits, and shifts in the labour market and broader economy. Local governments have faced increasing demands for public services often in a fiscal context of shrinking revenues and, especially in Southern California, even municipal bankruptcy.

Because of its importance in the changing public policy response to homelessness, the diversity of local governmental responses has encountered increasing exploration in scholarly research (e.g. Blau 1992; Stoner 1995; Wolch and Dear 1993). Such research has shown that municipal response is often inherently contradictory, with some programmes reflecting the desire to provide services and housing for homeless persons, and others reflecting more rejecting tendencies, such as the criminalization of homelessness and other NIMBY-type local legislation. The question of the NIMBY syndrome as expressed by the local state is quite complex, made so by the particularities of homelessness and municipal response. In this chapter, NIMBY is considered in relation to a specific issue: the criminalization of homelessness through anti-camping ordinances. At issue are the internal coherence of municipal government rejection, how anti-camping ordinances develop, and how local legislation may ultimately problematize the potential for accepting public response to social crises such as homelessness and HIV/AIDS. These issues are explored with respect to the recent (and ongoing) fight to ban camping in public spaces in the City of Santa Ana in Orange County, California. This dispute involved the City of Santa Ana, the County of Orange, homeless individuals, and Legal Aid Society and American Civil Liberties Union (ACLU) lawyers, and centred upon whether the arrests of homeless persons with camping paraphernalia were unconstitutional. The court battles over the Santa Ana anti-camping ordinance proved extremely important in constructing a legal precedent for developing and implementing local ordinances making homeless activities illegal and punishable. The court decisions upholding the Santa Ana ordinance paved the way for other cities not only in Southern California but across the nation to look to such legal strategies for removing the homeless 'problem' from their jurisdictions.

A recently released report from the National Law Center on Homelessness and Poverty estimated that at least forty-two cities nationwide have engaged in the enforcement of vagrancy and loitering ordinances or the development and implementation of new anti-camping and anti-panhandling laws (Anonymous 1995). The proliferation of local ordinances in Southern California and across the nation to remove homelessness from municipal boundaries is likely to continue given the persistent growth in the numbers and dispersion of homeless persons and the inability of higher levels of government to address the issue effectively. In addition, local ordinances banning activities or characteristics associated with homelessness have become popular as municipalities become unwilling to take on what they perceive to be an unfair burden of homeless service provision.

Municipal Response to Homelessness: From Providing Services to Legislating NIMBY

Homelessness is a growing issue over which local governments are coming into increasing conflict. Not only has homelessness become an increasingly significant issue across many cities and municipalities, but local governments have also had to adjust to a context of increasing balkanization of local political interests in large metropolitan areas. Fragmentation of metropolitan areas is not a new phenomenon, and indeed, many public choice theorists have lauded such developments as ways of allowing individuals and households to state their preferences in terms of public services and tax burdens (Tiebout 1956). This fragmentation of political interests and the protectionist tendencies that have sprung from such fragmentation have become particularly important in the local state's response to homelessness. Growing metropolitan fragmentation, inter-jurisdictional conflict, and socio-spatial segregation by race and class have been the result of middle- and upper-class household efforts to deflect the fiscal, social, and economic burdens of central city areas. In efforts to keep inner-city problems from entering middle- and upper-class enclaves, municipal governmental bodies have increasingly used legislation (e.g. local ordinances) to prevent the growth of local homeless populations, or to remove homeless persons from local jurisdictions. Thus, rather than regional solutions to homelessness, metropolitan fragmentation has conditioned a NIMBY-type response, to remove not only the 'problem' of homelessness from jurisdictional boundaries, but also the responsibility for providing services and programmes for homeless individuals and families.

Social Services at the Municipal Level

Municipalities have been increasingly pressured to respond to the growing prevalence and visibility of homelessness within their boundaries. Local agencies have been involved in varying ways in trying to address home-lessness, either through preventative means or in coping with the ongoing crises of homeless needs. When municipalities have been involved in the provision of services, the private, non-profit, and public sectors have participated in differing capacities. Joel Blau (1992) categorizes these varying forms of participation into three groups: (1) private, non-profit provision of services, with little or no municipal government participation (e.g. Miami and Houston); (2) the contracting out of services to the private, non-profit service sector by municipal government, many but not all in public–private partnerships (most large cities operate using this model, for example Chicago, San Francisco, Los Angeles, St Louis, and Seattle); and (3) municipal government operation and management of shelters, but also the contracting out of services to the local private, non-profit shelter and services sector (Philadelphia, Boston, and New York). Since the 1980s, private, non-profit agencies have provided the mainstay of homeless services (providing direct services, space, food, and labour), with private sector organizations (often religious and non-profit agencies) and, increasingly, federal funding sources providing financial support.

The context of changing federal and state legislation regarding homelessness has provided an important backdrop to the differentiated responses of local municipalities. The Stewart B. McKinney Homeless Assistance Act of 1987 is the most recent, and probably the most well-known, federal legislation in the US providing assistance for homeless persons, through direct support of emergency shelters, and appropriation of funds for physical and mental health care, substance abuse services, education and job training, and services for homeless veterans. Martha Burt and Barbara Cohen (1989) in their analysis of federal, state, and local programmes, however, also point to other federal programmes, which, during the 1980s, also were designed to provide aid to homeless persons primarily through food assistance: the Emergency Food and Shelter Program (EFSP) within the Jobs Stimulus Act of 1983, and the Temporary Emergency Food Assistance Program (TEFAP) authorized in 1983. But many of the beneficiaries of the food assistance programmes, particularly TEFAP, tended not to be homeless. In recognition of the lack of appropriateness of such legislation for homeless needs, Congress passed the Homeless Eligibility Clarification Act in 1986 (targeting Aid to Families with Dependent Children and Supplemental Security Income) to improve access of income maintenance programmes to homeless persons. In addition, the Hunger Prevention Act of 1988 was developed to address many of the mismatches encountered in the implementation of TEFAP by providing federal assistance directly to soup kitchens and shelters. Most states did not provide programmes or funding assistance for homeless services prior to the passage of the McKinney Act, but those that did have continued to outpace those states not providing services in terms of resource allocation to homelessness.

Within this context of institutional response at higher levels of governance, Blau (1992) argues that the reasons that municipalities differ in their response to homelessness stem from two primary sources, first, the visibility and prevalence of the homeless population within municipal boundaries, and, second, the composition of the coalition governing local politics. The visibility of the local homeless population can influence local politics towards helping or rejecting policies. If the growing homeless population appears more worthy, that is, they have become homeless because of economic downturns or natural disasters, then municipal response might include the provision of temporary shelter, food, clothing, and other material and emotional support. However, if the growing homeless population represents an expansion of those unwanted, unproductive, and stigmatized individuals often representing homelessness in the public consciousness, then municipal response is likely to be punitive and expulsive. The governing political coalition is the second factor which Blau (1992) argues conditions municipal response. The importance of this political network is that 'it is the coalition that monitors the risk of capital flight and determines whether the comparative costs and benefits of the city's policy will make businesses want to leave' (p. 131; also Mair 1986).

Municipal Responses to Homelessness

These issues become paramount in metropolitan areas with large populations of homeless persons. A recent study by Robin Law (1991) of the eighty-five cities in Los Angeles County both reinforces and expands Joel Blau's assertions. Law, mirroring Blau's research, found a wide range of policies enacted by municipalities in Los Angeles County in trying to address the critical and growing nature of homelessness; Los Angeles is a particularly important site for defining the local politics of homelessness, with its national identity as the 'homelessness capital' of the US (Wolch and Dear 1993). Law classified city responses in Los Angeles County to homelessness in terms of three dimensions: local spending on services associated with homelessness; social control practices, such as anti-camping ordinances and police sweeps of homeless encampments; and zoning, especially changes to zoning codes expressly associated with homeless services (such as shelters, soup kitchens, transitional housing and single room occupancy residences). The municipal responses varied from little or no action to aggressive policies, with some cities providing services to local homeless populations and others using police sweeps and municipal ordinances to remove homeless persons and encampments from public space. From a comparison of city resources, demographic composition, and municipal actions concerning homelessness, Law found no particular characteristics which could explain the variation in municipal response. Rather, like Blau, she surmised that municipal response was 'much more a function of political choices than simple municipal resources' (Law 1991: 22).

By and large, for most cities in Los Angeles and other regions across the US, the common municipal response to the growing need for housing and human services by homeless persons has been not to respond at all. Such political choices in regions experiencing growth and dispersion in the homeless population emanate from the history of service provision and political generosity or a lack thereof (which includes the history of municipal autonomy and the past and present provision of services for low-income and homeless individuals), class-based and turf-based politics concerning the negative outcomes of development (such as who should pay for the lack of affordable housing, traffic congestion, and environmental degradation), and perhaps most importantly, the diminishing sense of responsibility that municipalities feel towards care for homeless and indigent individuals (Wolch and Dear 1993).

   This diminishing sense of responsibility, and, increasingly, the outright rejection of local responsibility for homeless persons stems in part from what Law (1991) argues is 'the perceived "portability" of the homeless-ness problem' (p. 28). That is, there is the widespread perception that homelessness is not permanently tied to particular locations. By its very definition of a lack of a permanent place to sleep and live, homeless persons must continually search for services, shelter, and material and emotional support, requiring a minimum level of mobility to cope with daily needs. Thus, policy-makers tend to view homelessness as a problem which can be relocated, thereby eliminating the undesired presence of homeless persons and the need for provision of services. The fragmenta-tion and balkanization of political interests in large metropolitan areas like Los Angeles have exacerbated this practice of the expulsion of home-lessness. As Law (1991) argues, 'The cities that did provide relatively generous services to homeless people are vulnerable to becoming "suckers" in a metropolitan tragedy of the commons, incurring not only financial costs of service provision but also political costs of intense public contro-versy' (p. 29). Cities in Los Angeles County, such as Venice and Santa Monica, known widely as relatively generous in terms of policies concern-ing homelessness, have become sites of escalating tensions as residents and politicians have become increasingly frustrated over the growing numbers of homeless persons living in the area (Wolch and Dear 1993). The parti-cular sites of conflict often involve public facilities, such as parks and toilets, and other places where homeless persons congregate or panhandle.
   Although not all residents of these and other cities are hostile to home-less persons, cities have increasingly turned to anti-camping, anti-loitering, and anti-panhandling/anti-begging ordinances to respond to public frust-ration and mounting violence directed at homeless persons and human service facilities. As Mike Davis has argued, such controversies signify that 'there is no longer a single, universal standard of entitlement to the use of public space' (1991: 325). Instead, conflict over whether homeless persons have a right to sleep in or otherwise use public space signifies what Don Mitchell (1995) has argued are the opposing and perhaps insurmountable visions of the nature of public space. In his exploration of the conflict over People's Park in Berkeley, California, Mitchell succinctly encapsulates these 'irreconcilable, ideological visions' about the appropriate nature and use of public space:

Activists and the homeless people who used the Park promoted a vision of a space marked by free interaction and the absence of coercion by powerful institutions. For them, public space was an unconstrained space within which political move-ments can organize and expand into wider arenas. . . . The vision of representatives of the University (not to mention planners in many cities) was quite different. Theirs was one of open space for recreation and entertainment, subject to usage by an appropriate public that is allowed in. Public space thus constituted a controlled and orderly *retreat* where a properly behaved public might experience the spectacle of the city. (p. 115; emphasis in text)

Criminalizing Homelessness[2]

Although municipalities are increasingly turning to local ordinances as means of relocating their homeless populations, such legal strategies which would remove or prevent undesired persons from entering or remaining in cities and neighbourhoods are by no means new. Research on legislation concerning low-income and indigent groups has long described the construction and implementation of Elizabethan Poor Laws in England, working both to control labour and to minimize criminal behaviour.

Legal Arguments of Vagueness and Discrimination

More recently, researchers have cited early 1970s court decisions in the US at various levels of the legal system where legal advocates have argued for overturning vagrancy ordinances and for declaring the existence of a 'right to shelter'.[3] In the landmark *Papachristou* v. *City of Jacksonville* of 1972, the City of Jacksonville's vagrancy ordinance was challenged in the State of Florida Supreme Court by eight people who were convicted under this ordinance. In a unanimous decision, the Court overturned the ordinance on due process grounds, stating that the ordinance was vague, and there-fore 'failed to give fair notice of prohibited conduct' and 'encouraged arbitrary arrests and convictions' (Stoner 1995: 24). The US Supreme Court has also acted to overturn local anti-vagrancy and anti-loitering ordinances, challenged by legal advocates for homeless persons. Similar concerns about vagueness were expressed by the Court in its striking down of a California loitering statute in *Kolender* v. *Lawson* in 1983. This local statute had 'punished failure by any person wandering the street to produce credible identification when requested to do so by a police officer' (Stoner 1995: 24).

Federal courts have also been active in overturning state laws making panhandling or begging illegal. Using similar arguments about vagueness and the suppression of First Amendment rights, lawyers successfully argued in 1991 and 1992 against laws in New York (*N.Y. Police Dept.* v. *Loper*, 90–Civ. 7545–S.D.N.Y.) and California (*Blair* v. *Shanahan*, 775 F. Supp. 1315, N.D. Cal. 1991). Several lower courts during the early 1980s expressed similar views about the vagueness and discriminatory impact of local anti-vagrancy and anti-loitering ordinances. A municipal court in Carmel, California, for example, in 1971 overturned a vagrancy ordinance which had been enacted in 1968 to 'prohibit "hippies" from sitting on sidewalks, or steps, and sitting or lying on lawns' (Stoner 1995: 158). In

Parr v. *Municipal Court for Monterey-Carmel, J.D.* of 1971, the Court of
Appeal struck down the ordinance, arguing that 'the discriminatory impact
is achieved by the hostile tone and the critical description directed to one
segment of society' (479 P.2d 353, 1971 quoted in Stoner 1995: 158). A
similar lower court decision was found in Pompano Beach, Florida, in 1984
(*City of Pompano Beach* v. *Capalbo*, 455 So. 2d 468, 269–70, Fla. Dist. Ct.
App. 1984).

In general, however, lower court decisions concerning the vagueness of
municipal ordinances have been much more mixed. As Madeleine Stoner
(1995) has argued, such decisions have indicated 'the reluctance of the
lower courts to overturn vagrancy ordinances' developed by municipalities
to control a growing homeless population (p. 156). While there have been
instances (such as *Rubin et al.* v. *City of Santa Monica* in 1993)[4] where
lower courts found that anti-vagrancy and anti-loitering ordinances were
unconstitutional, based on First Amendment rights, many more decisions
have supported the power of local municipalities to develop ordinances
banning vagrancy, loitering, sleeping, or camping. In Phoenix, Arizona,
San Francisco, and Santa Barbara, California, local ordinances banning
sitting, lying, or sleeping in public spaces, in public parks, or public right of
ways have been upheld by lower courts and Courts of Appeal rejecting
First and Fourth Amendment rights and challenges of vagueness (*Seeley* v.
State of Arizona, 655 P.2d 803, 1982; *Stone* v. *Agnos,* F2d, 9th Cir. 1992, 92
Daily Journal D.A.R. 4524, No. 91–15206, April 3, 1992; *People* v.
Davenport, 222 Cal. Rptr. 736, Sup. Ct. App. Civ. 1985, *cert. denied,* 475
U.S. 1141, 1986 cited in Stoner 1995: 158).

Arguments Citing Cruel and Unusual Punishment

Other challenges have sought to deem anti-camping ordinances as cruel
and unusual punishment based solely on persons being homeless (i.e.
punishment for status). Applying the 'Robinson Doctrine' (based on
the 1962 case *Robinson* v. *California*) would void such ordinances on the
grounds that they criminalize status (Smith 1996). Challenges of this type
have taken place in the federal district court (challenging municipal laws in
the City of Dallas, the City of Miami, and the County of San Francisco)
and state appellate court levels (*Tobe* v. *City of Santa Ana*). As Smith
(1996) argues, 'the success of the [Robinson Doctrine] argument depends
on the extension of Robinson to involuntary acts that are symptomatic of a
protected status' (p. 319). Such arguments have proved to be moderately
successful; of the four cases published and analysed by Smith (1996), two
(*Johnson* v. *City of Dallas* and *Tobe* v. *City of Santa Ana*) were later overturned for reasons other than whether the ordinances violated the Eighth
Amendment (i.e. cruel and unusual punishment). In the City of Miami, the
court ordered the City to cease its arrests of homeless persons sleeping in

two public areas, and in the City of San Francisco, 'the court directly rejected the Robinson Doctrine argument "as applied" to the homeless' (Smith 1996: 325). I will return to the case of the anti-camping ordinance in the City of Santa Ana later in the chapter.

Anti-Homeless Laws: the Domino Effect

In a recent study of sixteen cities across the US, the National Law Center on Homelessness and Poverty (1993) found that 80 laws had been enacted banning begging, sleeping, camping, loitering and destruction of property (cited in Stoner 1995). Of these 80 laws, 25 were local ordinances specifically directed at sleeping, camping, loitering, or sitting in public spaces or parks. In Southern California, between 1992 and 1993, eight cities passed ordinances banning sleeping or camping in public spaces (Stoner 1995). Stoner suggests that this group of ordinances enacted in such a short time span represents 'a domino-effect in which neighboring communities copy each other' (p. 151). She argues that this domino-effect has occurred because municipalities fear that homeless persons will be evicted from neighbouring places only to congregate within their jurisdiction and in addition that there are widely held perceptions of the possible responses to homelessness (i.e. that municipal agencies tend to think in similar ways about such problems). Beyond these two reasons, there is also the context of concomitant court decisions, which at least in the lower courts, seemed to be favouring the municipal authority to legislate against unwanted individuals (e.g. *Stone* v. *Agnos*, 1992, in San Francisco).

The cities in Southern California passing such ordinances (Long Beach, Santa Monica, Fullerton, Santa Ana, Orange, Santa Barbara, and Beverly Hills) tended to base them on the model developed by the City of West Hollywood which defined camping in a very specific way to avoid legal challenges. The West Hollywood ordinance prohibits camping which is defined as 'residing in or using a park for living accommodation purposes, as exemplified by remaining for prolonged or repetitious periods . . . with one's personal possessions' (Hill-Holtzman 1992: B2). The laws in West Hollywood and Santa Monica were both written by the law firm of Richards, Watson & Gershon, and were 'declared legal by a three-member appellate panel of the Los Angeles County Superior Court' (Hill-Holtzman 1994: J3). The City of Santa Ana's ordinance represented a particularly important case, as it was challenged by advocates, homeless persons, and Legal Aid Society and ACLU lawyers in the California Supreme Court. The court decisions made in this case would determine to a large degree whether these other cities would be able to implement their ordinances, and, more importantly, whether other cities across the US would propose and implement similar ordinances in the future.

The Anti-Camping Ordinance in Santa Ana

The Santa Ana anti-camping ordinance was the result of a series of events and legal confrontations among legal advocates, City officials, differing levels of the court, the police, and homeless individuals. Although the ordinance has often been viewed solely from a legalistic point of view as the culmination of precedents and legal battles, the development of the ordinance passing judicial muster was the result of trial and error, individual personalities, and a municipal context where the number and behaviour of homeless persons had tipped the scales from tolerable to unacceptable for particular individuals with the power and influence to criminalize and evict homeless persons from the Civic Center plaza area.

Sowing the Seeds of Conflict

Although the ordinance was passed by the City in 1992, the seeds of this ordinance were sown earlier in the decade. The City had been pursuing a downtown revitalization strategy to encourage business investment and economic development. As part of this strategy, the Santa Ana Community Redevelopment Agency had worked to develop office complexes (such as 801 Civic Center, costing $60 million and built in 1983) in the downtown area, near the County and City seats of government and the main branches of the County's superior and municipal courts. The City's Redevelopment Agency hoped that new construction might spark a rejuvenation of downtown commerce.

Such economic development efforts at business growth were seen to be at odds with the growing concentration of homeless persons in the City's Civic Center area. In 1988, the countywide homeless population was estimated to be between 5,000 to 6,000 persons, but many persons believed that about half of them spent 'most of their time in Santa Ana' (Schwartz and Kurtzman 1988: I3). In June 1988, the City of Santa Ana's maintenance crews began implementing a City policy to confiscate unattended personal property which was stored on public property. Elected officials were growing weary and frustrated over the seeming unequal burden that had been taken on by the City because it happened to host more services and housing for homeless persons. That is, as a consequence of the City's proactive strategies in providing services as compared to neighbouring municipalities, it was having to shoulder a greater municipal share of a growing homeless population. Mayor Dan Young in 1988 argued that 'Santa Ana should not be asked to accommodate the majority of the county's homeless' (Schwartz and Kurtzman 1988: I32).

In the same year, the City of Santa Ana and its police department came into direct conflict with advocate lawyers from the local Legal Aid Society and the ACLU over the City's attempts to disperse the homeless population through police sweeps and confiscation of private possessions (e.g. sleeping bags and other private property). The eventual lawsuit brought by Legal Aid Society and ACLU lawyers was later settled in August 1991, when the City Council voted (5–2) to settle the lawsuit brought in 1988 concerning the police sweeps. While this lawsuit was brought to the courts and settled in 1991, the City was working to develop alternative means to rid itself effectively of its homelessness 'problem'.

Perhaps as a consequence of the ongoing public and elected official frustration over the greater responsibility shouldered by the City, in the summer of 1990, Santa Ana police arrested 64 homeless persons on charges including jaywalking, littering, public drunkenness, and urinating in public (Anonymous 1991; Martinez 1991*a*). Arrested individuals were 'chained to benches at [Santa Ana Municipal Stadium] for up to six hours' while they were processed, cited, and released (Anonymous 1991). Protests against these arrests were made by homeless advocates and Latino spokespersons, but the arrests continued, with 26 additional persons arrested a week later (Anonymous 1990). Santa Ana Police Chief Paul Walters argued that 'the crackdown is in response to disorderly behavior in the Civic Center, and that, if left unchallenged, it will send a signal that invites more anti-social behavior' (Anonymous 1990).

These activities were highly publicized, and, in July 1991, a group of homeless persons sued the City, claiming that their civil rights had been violated. Lawyers from the Legal Aid Society and ACLU brought the case to Santa Ana Municipal Court on behalf of this group. After the City moved to prosecute the arrested homeless persons, Municipal Judge B. Tam Nomoto 'threw the case out and declared the city's actions illegal and discriminatory' (Martinez 1991*a*: A21). The municipal court found that the police had 'deliberately and intentionally implemented a program . . . targeting the homeless' (Goldberg 1994: 102). Thirty-one homeless plaintiffs won a $400,000 settlement from the City of Santa Ana, after suing the Santa Ana Police Department for their arrests during police sweeps in the Civic Center complex (Martinez 1991*b*). The City chose to settle for $400,000, since, as City Attorney Edward Cooper argued, fighting the case in court would have cost the City more than $450,000, or up to $15,000 for each case.

The First Anti-Camping Ordinance

But City officials did not give up on pursuing legal avenues for relocating the homeless population from the Civic Center. In August 1992, frustrated by the growing presence of homeless persons, City officials passed the first of their multiple attempts at a local anti-camping ordinance (Sachs 1993: 38). As Orange County Senior Deputy District Attorney E. Thomas Dunn Jr. argued, 'You would think you were in the middle of a sewer plant', describing the encampment of lean-tos and campfires seen from Civic Center offices (Dolan 1994, A37). Indeed, much of the impetus for creating the ordinance came from Civic Center offices (housing County and local government agencies, and superior and municipal courts), whose workers were frustrated by the seeming permanence of the homeless population. The anti-camping ordinance, which went into effect in September 1992, made 'it a crime punishable by up to six months in jail to use a sleeping bag or blanket or to store personal effects on public sidewalks, streets, parking lots and government malls' (Dolan 1995: A1). In addition, this ordinance stated that homeless persons 'were to clear out of town by sunset' (Di Rado 1994: A27). Police in Santa Ana began systematic sweeps of the Civic Center plaza area, warning homeless persons that sleeping bags, blankets, and other camping paraphernalia could not 'be used to guard against the elements' (Gurza 1992: B8). Police officers tried to ensure that homeless persons had been properly informed about the ordinance by videotaping and photographing their distribution of fliers prior to enforcement (Shaffer 1992). Having camping paraphernalia was not illegal, only using it to sleep in the Civic Center. According to the Deputy City Attorney in Santa Ana, Paul Coble, sleeping in the Civic Center without bedding or blankets remained a legal activity (Gurza 1992).

E. Thomas Dunn, the County's Senior Deputy District Attorney, argued that 'the city ought to have a right to keep [public areas] clean for the rest of us. There's no reason why there ought to be a Republic of the Homeless, like an abdication of a piece of territory to those who are above the law' (Romney 1994: A25). However, he also argued that this ordinance would not be used against those who had no alternatives to living on the streets; instead such individuals 'would have a legitimate legal defense in court against the ordinance and would be spared punishment' (Dolan 1994: A37). In addition, Police Chief Paul Walters argued that police sweeps of the Civic Center were responses to complaints made about 'unsanitary and unpleasant conditions associated with the homeless' (Gurza 1992: B8).

In September 1992, ACLU, Legal Aid Society, and National Lawyers Guild attorneys filed suits against five cities primarily in Orange County (Fullerton, Long Beach, Orange, Santa Ana, and Santa Barbara), challenging municipal ordinances directed against homeless persons camping on public property. These challenges argued that the local ordinances subjected homeless persons to cruel and unusual punishment and violated their constitutional rights of travel and movement (Frank 1992).[5] In Santa Ana, Legal Aid Society lawyers filed suit against the City in Orange County Superior Court. The City had changed its tactics from the visible and confrontational mass arrests of 1990 to documented warnings and confiscation of evidence before citing violators of the ordinance. Harry Simon, of the Legal Aid Society of Orange County, argued that this new set of strategies emanated from the judicial decision to award homeless plaintiffs $400,000; 'That was quite an expensive event for them. I think they'd prefer to take a more low-key approach' (Shaffer 1992: B1). Beginning 21 October 1992, police officers began to warn homeless persons in the Civic Center that, from 27 October, 'people faced arrest and confiscation of their property if they did not voluntarily comply with the city's ban against camping in public places' (Gurza 1992: B8). Once enforcement of the ordinance began, police officers would begin to seize personal property (e.g. blankets, shopping carts, and tents) as evidence in prosecuting violators. In April 1993, Orange County Superior Court Judge James L. Smith, in deciding a legal challenge to the constitutionality of the anti-camping law, upheld the ordinance, rejecting arguments made by Legal Aid Society lawyers that the ordinance was vague and constituted cruel and unusual punishment.

By late June 1993, Central Orange County Municipal Court Judge Gregory Lewis had expressed concerns over the potential clogs in the court calendar caused by homeless litigation. Santa Ana police had issued 90 citations since January 1993 to homeless persons violating the ordinance; however, the cases had yet to go to trial. Twenty-five homeless defendants had pleaded innocent to charges of violating the City's anti-camping ordinance, with expected jury trials on 2 August. The municipal court at the time had an additional 40 court cases pending trial. Because of the concern over clogging an over-burdened court system, the District Attorney's office had charged 53 defendants, but delayed the charging of 40 other homeless individuals (Chow 1993). The 25 defendants scheduled to stand trial on 2 August had been among the first individuals cited for misdemeanour violation of the anti-camping ordinance, but their arraignments were postponed pending a decision in civil court concerning the constitutionality of the law. Legal Aid Society attorneys proposed to the City that they would end the litigation in exchange for the City allowing homeless persons to use blankets and bedding at night, and to store their personal property during the day. The City rejected the proposal, with City Attorney Edward Cooper arguing that the offer was 'not a proposal worth considering' (Gurza 1993a: B1).

In July 1993, a three-judge panel of the 4th District Court of Appeal in Santa Ana granted a temporary injunction preventing the City from implementing its anti-camping ban until constitutional issues could be decided (Gurza 1993b). The State Court of Appeal reversed Orange County Superior Court Judge Smith's prior decision to uphold the ordinance, halting the cases of the 25 homeless persons who were scheduled for trial on 2 August (Gurza 1993c). In making its ruling, the three-judge appellate panel ruled that the ordinance was 'vague, imposes class-based restrictions on the ability to live and travel, and constitutes cruel and unusual punishment' (Di Rado 1994: A27).

The Second Anti-Camping Ordinance

Two weeks after this ruling, the City continued to expel homeless persons from the Civic Center using an 1872 state anti-vagrancy law banning lodging on property without express permission by the owner. The 121-year-old state law had originally been used to evict vagrants from barns, sheds, and outhouses, and had been amended in the mid-1960s. Legal advocates, however, successfully obtained a temporary restraining order in October barring the City from enforcing this ordinance until 22 October when the Orange County Superior Court would hear lengthier arguments and decide on whether a longer injunction was warranted. The primary argument made by Legal Aid Society attorneys centred on the word, 'lodging'. Harry Simon, of the Legal Aid Society, argued that, in the Civic Center plaza, homeless persons slept outside, and therefore were not in violation of the nineteenth-century vagrancy ordinance (Weston 1993).

After the temporary restraining order was issued on 8 October 1993, the City protested, challenging the decision to impose the restraining order, and presented videotapes and photographs illustrating the conditions at the plaza. This challenge marked the first time that the City used evidence to try to support their argument about the necessity of the ordinance. Orange County Superior Court Judge Robert J. Polis upheld his earlier restraining order, deciding that individuals could be cited with a misdemeanour offence for illegally occupying buildings (such as parking garages), but that homeless persons could 'remain unencumbered in the center's plaza, parking lots and sidewalks' (MacWilliams 1993: B1).

The City continued to pursue the development of a local ordinance even after the Appeals Court decisions on the City's first two attempts at constructing and implementing an anti-camping law. To overcome the problem of vagueness, the issue of travel restrictions, and the interpretation that the prior ordinance constituted cruel and unusual punishment, the City's third attempt at an ordinance was based on an existing US Park Service law (upheld by the US Supreme Court), which banned camping and was used to relocate homeless persons in Lafayette Park, Washington, DC (Santoyo 1993*a*). In addition, while previous anti-camping ordinances applied citywide, the newest incarnation of the ordinance would be enforced only in the Civic Center. Harry Simon, of the Legal Aid Society, responded to this newest attempt at constructing a legal ordinance, 'I'm getting tired of filing these lawsuits, and I would hope that the city would get tired of losing them' (Gurza 1993*d*: B7).

During these legal battles, the presence of homeless persons continued to frustrate County, City, and municipal court employees. As County planner

Leon Kolankiewicz argued, 'I would find my sympathy wearing out when there is a welcome mat that smells and is covered in flies. It was unbecoming of the county seat of one of the major counties in the country' (Santoyo 1993*b*: 1).

The Third Time is the Charm: Santa Ana Successfully Designs an Anti-Camping Ordinance

The development of the third anti-camping ordinance occurred during the same period that the County began massive budget cuts in response to the largest municipal bankruptcy case in US history. The bankruptcy, which resulted from massive losses emanating from a risky municipal investment pool caused service cutbacks across departments and services, and translated into direct service and monetary support cutbacks for Orange County's low-income residents (Moehringer 1994). Of the $40.2 million in budget reductions approved by the County Board of Supervisors across twenty-seven County departments and programmes, 40 per cent or $16.6 million had supported health care, social services, and defence attorneys for the County's low-income residents (Weikel and Marquis 1994). Rent assistance was cancelled, and job training programmes and several programmes for mentally disabled individuals were eliminated. Overall, $3 million was cut from the Health Care Agency (providing public health, medical, and mental health services), $8.8 million was cut from the Social Services Agency (administering welfare, foster care, and other benefits), and $12.7 million was cut from the alternative defence budget (providing private defence attorneys for low-income adults and juveniles). The County's Co-ordinator of Homeless Services was one of the first positions eliminated by the budget cutbacks. The thirty to fifty non-profit agencies providing services to homeless persons no longer received County funding after the bankruptcy, including the organization which was contracted to run the emergency shelters in the national armouries (providing about 300 men, women, and children with beds during the winter season).

The budget cutbacks exacerbated the crisis of poverty and homelessness in the County and in the City of Santa Ana. Human services in Orange County had already failed to keep pace with the growing population of homeless persons. Advocates have argued that there are fewer than 1,000 shelter beds available for the 12,000 to 15,000 persons who are homeless on any given night (Dolan 1995).[6] In particular, recent estimates indicate that the City of Santa Ana has approximately 3,000 homeless persons on any given night and only 322 shelter beds. One of the many reasons for the lack of adequate homeless services and shelter, and the substantial cuts made after the municipal bankruptcy, was the lack of public prioritization of homelessness as an issue requiring policy intervention. Although poverty and homelessness have become increasingly prevalent in Orange County, there has not been a concomitant public recognition that these issues constitute a high priority. According to public opinion polls of Orange County residents, high priority public policy issues include crime, transportation, and the economy (Baldassare and Katz 1994). Such public opinion polls indicating that homelessness is not a high priority among residents belies the local public policy context concerning homelessness, where there has been intense community opposition to human services associated with homelessness and, especially in Santa Ana, ongoing attempts to legislate homeless persons out of the jurisdiction.

January 1994 brought another challenge by Legal Aid Society attorneys to block the implementation of the City's third anti-camping ordinance. This time, Orange County Superior Court Judge Robert J. Polis refused to issue a restraining order blocking the ordinance, arguing that the newest ordinance would probably stand up to constitutional challenges (Volzke 1994*a*). But a month later, the 4th District Court of Appeal (in early February 1994) struck down the City's 1992 anti-camping ordinance, calling the ordinance 'constitutionally repugnant' after presentation of internal memos indicating that the City operated from a 'hidden policy' working to force homeless persons from City boundaries (Miller 1994). In a memo dated 16 June 1988 and cited by the Appeals Court, Parks Director Allen E. Doby, wrote: 'City Council has developed a policy that the vagrants are no longer welcome in the city of Santa Ana' (Gurza 1994: A2). The ruling also commented on the third and newest ordinance, also facing a legal challenge, writing that the City 'apparently thinks the courts will perceive naught but a benign motive. Against the history of the city's war on its weakest citizens, that is asking a lot' (Gurza 1994: A2).

However, in May 1994, the California State Supreme Court agreed to review this appellate decision, nullifying the Appeal Court's overturning of the law. Perhaps as a foreshadowing of the Supreme Court decision to be handed down at the end of the year, in September 1994 Orange County Superior Court Judge Robert J. Polis heard arguments challenging the latest incarnation of the City's anti-camping ban, and upheld the ordinance, disagreeing with Legal Aid Society and civil rights attorneys' arguments that the law violated homeless persons' rights to travel and punished them for their status as homeless. In his ruling, Judge Polis argued that while the City had used 'warlike' tactics in 1988, it had recently become more 'sensitive' in confronting the homeless population in the Civic Center (Volzke 1994*b*: B1). In the autumn of 1994, the first Latino mayor ever elected in Santa Ana, Miguel Pulido, took office and continued the City's strategies of moving the homeless population from the Civic Center.

The State Supreme Court after hearing arguments handed down its decision about the anti-camping ordinance in December 1994. In a hearing in San Francisco, the ordinance was challenged 'on the grounds that it violated the right to travel, punished people on the basis of their status and was unconstitutionally vague' (Dolan 1995: A10). The City of Santa Ana (represented by Santa Ana Assistant City Attorney Robert Wheeler in front of the state's highest court) argued that the appellate decision which had overturned the prior anti-camping ordinance had 'tied cities' hands in their efforts to govern and wrongly created a constitutional right to live on public property' (Romney 1994: A1). This 1994 lawsuit constituted the first time that the constitutionality of anti-camping laws had been raised at this level of the court. Attorneys for both the City and for the homeless plaintiffs predicted that this decision, if it upheld the Santa Ana ordinance, would 'encourage more cities to pass laws limiting homeless people's access to public property' (Dolan 1994: A37). Indeed, at the time of the Supreme Court decision regarding the Santa Ana ordinance, five other cities across the nation (Miami, Dallas, Baltimore, San Francisco, and Seattle) had all become embroiled in lawsuits concerning local ordinances regarding homeless persons (Romney 1994).

The importance of this judicial decision was also clear by the number of attorneys, justices, and law professors supporting either side. On the side of the homeless plaintiffs, supporters of the lawsuit against the City of Santa Ana included the US attorney general, retired State Supreme Court justices, and constitutional law professors from several of the country's most prestigious universities. The US Department of Justice, Civil Rights Division, filed a 24-page brief, stating that the Santa Ana anti-camping ordinance violated the Eighth Amendment by 'criminalizing homeless persons' sleeping in public even when no shelter is available' (Romney 1994: A25). On the side of the City of Santa Ana were more than ninety city attorneys from throughout the State of California (drafting a friend of the court brief) and conservative legal foundations (such as the Pacific Legal Foundation, the Criminal Justice Legal Foundation, and the American Alliance for Rights and Responsibilities) (Romney 1994).

In its decision on this case, the California Supreme Court in April 1995 in a 6–1 decision upheld Santa Ana's anti-camping ordinance, holding that it did not violate the constitutional rights of homeless persons (Dolan 1995). This decision was the first by any state supreme court on a local ordinance seeking to prosecute activities or behaviours associated with homelessness. In the majority written decision, Justice Marvin Baxter found that 'There is no right to use public property for living accommodations or for storage of personal possessions, except insofar as the government permits such use by ordinance or regulation' (Dolan 1995: A10). Also in its ruling, the court declared that the criminal charges against a dozen homeless persons 'had been improperly dismissed' (Gurza 1995: A4). In addition, the court decision indicated that the obligation to provide services and shelter belonged to the County, rather than to the City. This position echoed arguments made by Santa Ana Mayor Miguel Pulido, that the City had neither the responsibility nor the resources to provide social services. Mayor Pulido believed that this decision was important not only for Santa Ana, but for other cities pursuing similar strategies to confront homelessness: 'When the ordinance was upheld by the Supreme Court, that was a big day for a lot of cities who are trying to regulate public space' (Hernandez 1996*b*: B5).

The drama did not end with the Supreme Court decision. In June 1995, Central Orange County Municipal Court heard arguments by attorneys representing seventy-three homeless persons cited for violating the anti-camping ordinance, that the charges should be dismissed because of the prohibitive costs to the Court of prosecuting each defendant. One of the lawyers argued that expert witnesses would cost the Court $25,000 over the $9,000 taxpayer cost for each day of the trial (MacWilliams 1995*a*). Attorneys for the homeless defendants recommended community service for those violating the City's anti-camping ordinance. The Court ruled that forty-nine of the homeless defendants would be allowed to complete eight hours of community service to the City by early August in exchange for a dismissal of the misdemeanour charges. Failure to complete their community service, however, would result in bench warrants for their arrests, resulting in a possible year in the county jail and a $1,000 fine. Lawyers argued that the difficult step in implementing this judgment would be locating the homeless defendants (MacWilliams 1995*b*).

Although most of the homeless defendants pleaded guilty, or had their cases dismissed because of a lack of a speedy trial, one homeless man, James Eichorn of Santa Ana, decided to challenge the anti-camping ordinance in May 1996. He was the first individual to go to trial, waiving his right to a speedy trial. As Eichorn, 49, argued, 'I believe it's a just cause. You can't push homeless people around just "cause they're homeless"' (Hernandez 1996*c*: B1). Eichorn did not have a history of alcohol or other substance abuse, and was a Vietnam War veteran. He lost his one-bedroom apartment when he was laid off from his machinist position in the early 1980s during an industry downturn. During the 1980s, he held various types of temporary employment, from a thrift-shop donations collector to a swap-meet attendant. From the wages earned from these jobs, Eichorn was able to live in hotels; however, since then, many of the cheap hotels in Santa Ana had been demolished or closed. Eichorn had since used his wages earned from day labour and giving blood for bus fares to travel to Fullerton and Anaheim several times a month to visit unemployment centres, and intended to save enough money eventually to buy a van so that he could 'drive around to find more and better-paying jobs' (Hernandez 1996*c*: B1).

The challenge brought by Eichorn and his attorneys, Todd Green and Brett Williamson, hinged on the argument that Eichorn had no choice but to violate the ordinance in order to survive; thus, citing Eichorn within the auspices of the ordinance punished him for his status of being homeless rather than illegal conduct. The premiss was that Eichorn was involuntarily homeless; therefore, he had no choice but to violate the ordinance. Police Lt. Bob Helton, however, argued that this was not the case: 'We're not arresting these people for being homeless, but because they're camping' (Hernandez 1996*a*: B1).

In a two-day, non-jury trial, Central Orange County Municipal Court Judge James Brooks in early May 1996 supported the City's position and rejected the arguments by Eichorn and his attorneys, stating that 'The evidence is overwhelming that he violated the law. Homelessness has nothing to do with it. He wasn't being cited for being homeless or being 126th in a 125-bed shelter. He was cited for camping in the city of Santa Ana in violation of the law' (Hernandez 1996*b*: B5). Eichorn was sentenced to forty hours of community service, but the sentence was suspended in expectation of an appeal to the judicial decision. Judge Brooks argued that Eichorn 'could simply apply for welfare or get help from his mother in Long Beach or brother in Connecticut' (Hernandez 1996*c*: B1).

Conclusion

Are local ordinances criminalizing the activities of homeless persons the wave of the future in terms of municipal response? On the basis of the case presented in this chapter, it might be claimed that in the wake of ongoing reductions in social services and welfare by the federal and state levels of government, we could only expect municipalities to follow suit. For municipalities, many of which are balancing multiple interests in often stringent fiscal circumstances, the relocation of homeless populations might seem the desirable solution in a homelessness crisis where federal and state public policy seems to have had few substantive effects. Abdication of social responsibility for those less fortunate appears to have become the popular choice in an arena where municipalities which do provide services and shelter are 'vulnerable to becoming "suckers" in a metropolitan tragedy of the commons, incurring not only financial costs of service provision but also political costs of intense public controversy' (Law 1991: 29).

And cities are less willing to be those 'suckers' in ongoing metropolitan conflict over delineating responsibility for a growing and diversifying homeless population. In a *Washington Post* column in 1995, George Will argued that the Supreme Court's upholding of the Santa Ana's anti-camping ordinance gave public agencies and officials 'some credit for being able to construe particular terms in reasonable contexts' (p. C7). In a critique of Legal Aid Society and other lawyers participating as what he termed 'gladiators for liberation' (quoting Karl Zinsmeister, the editor of the

American Enterprise magazine), Will closed with this comment, 'It speaks volumes about the country's condition that this elemental proposition had to wage a last-ditch fight for affirmation in the largest state's highest court' (p. C7).

The proposition was indeed elemental. The upholding of this ordinance in the state's highest court has provided legal precedent for other cities across the nation to engage in similar strategies to rid themselves of an ever-growing and diversifying homeless population. Prior to this court decision, forty-two cities across the US had already passed ordinances banning camping, panhandling, or loitering, or had begun enforcing vagrancy ordinances long in their municipal codes (but which had not been systematically or usually enforced). Advocates for both sides in the Santa Ana litigation predicted that there would be many more cities developing local ordinances in the wake of this Supreme Court decision.

If such ordinances continue to be upheld by the judicial system, what are some of the implications for understanding the NIMBY syndrome? At the outset, it is clear that such ordinances represent a legalization of exclusionary planning. To implement anti-camping ordinances, cities will be involved in ongoing sweeps of specific sites (such as the Civic Center area in Santa Ana), not necessarily targeting homeless persons *per se*, but citing individuals using camping paraphernalia in designated public spaces. Practically, this means that persons who are engaged in more appropriate or socially acceptable behaviour (such as picnicking or relaxing during work breaks), which, as George Will argued, can be construed as reasonable, will not be cited, or that persons without blankets or sleeping bags camping in designated areas would not be violating such ordinances. However, sleeping outside at night even in Southern California brings with it the potential for hypothermia and other health risks, necessitating protection from the elements. Effectively, homeless persons will have to find some place else to sleep. Since the Supreme Court decision, some homeless persons in Santa Ana have begun camping behind the offices of the Legal Aid Society, ostensibly not illegally since this is private property.

In the case of Santa Ana, the homeless population has also dispersed into nearby neighbourhoods, river beds, and private property (e.g. the offices of the Legal Aid Society). Residents in some of these areas have complained about the presence of homeless persons, using many of the same arguments workers in the Civic Center used prior to the design and implementation of the anti-camping ordinance. The point is that the issues of homelessness remain, and that anti-camping ordinances serve to move the homeless population through jurisdictions in search of a safer, although often temporary, location. Since many cities in Orange County are following the example laid by Santa Ana (i.e. passing local ordinances and implementing them while battling litigation in court), the possibilities for such locations shrink constantly. Legal Aid Society and civil rights attorneys continue to dispute the Santa Ana anti-camping ordinance in the courts, but the battle for the rights of homeless persons to remain in public space seems bleak.

The Santa Ana anti-camping ordinance has very important implications for the future of homelessness, particularly at the local level. As a municipal ordinance, it institutionalizes a NIMBY response, making available at any time the possibility of relocating homeless persons from a designated area (e.g. the Civic Center). In this case of NIMBY, the result does not reflect a response from residents and communities, but rather from the local state (here represented by the City of Santa Ana, and County municipal and court employees working in the Civic Center office buildings). They thus represent the interests of workers rather than residents, where homeless persons and homelessness more generally began to impinge on the daily routines of government agencies and officials. Indeed, one might argue that the ordinance was only possible because the homeless population was concentrated in an area where local state agents had to respond to their presence, rather than having to respond to a general issue of homelessness or to rejecting actions by residents.

Of course, municipalities have expressed NIMBY-type responses over homeless persons and other undesirable populations in the past. For example, large urban municipalities in California have informally transferred troubled adolescents to more remote cities and communities in the central and northern parts of the state to remove potential negative peer influences (both on the adolescents in question and on their peers) and to provide them with alternative physical environments in which rehabilitation could take place. Another common example has been the informal 'keep out' sign, through the lack of construction of human services and shelter for low-income and homeless persons, even with state laws mandating their construction and encumbering funds for this purpose. But the Santa Ana anti-camping ordinance goes beyond these informal mechanisms at removing and keeping out undesirable populations. The ordinance has close parallels with anti-vagrancy laws deemed unconstitutional in the early 1970s, which sought similar results but were declared vague and were argued to allow arbitrary citations and arrests. This new round of municipal ordinances in the 1990s, however, has passed judicial muster even at the highest state court, opening the way for a new group of anti-vagrancy ordinances to reappear in the lexicon of municipal policies for dealing with homelessness.

9

Intergovernmental Strategies to Reduce Stigma

HIV/AIDS Education and Prevention

FROM the data presented in this book, it is clear that HIV/AIDS is largely seen as a negative, and even dangerous, condition. Although medical research continues to seek a better understanding of the biology of HIV/AIDS, and researchers work to develop medical interventions both to treat and to prevent HIV/AIDS, there remains the widespread public fear of contagion. To address unwarranted fears and pervasive myths both about the nature of HIV/AIDS and its modes of transmission, public agencies and non-profit organizations have developed and used public awareness campaigns, and education and prevention strategies. These strategies during the 1980s were largely top-down, emanating from the federal level, primarily through the Centers for Disease Control and Prevention (CDC). States, localities, communities, and affected individuals (such as individuals living with HIV/AIDS and people not diagnosed HIV-positive but who are designated at high risk of HIV transmission) were largely excluded from this process. The CDC in 1993 in response to the growing challenges faced by public health agencies and service providers substantially altered these top-down practices and instituted a community planning requirement as a component of federal funding. This changing process has sought not only to educate the public about the modes of transmission of HIV/AIDS and to develop effective ways to curb transmission, but also to demystify the nature of HIV/AIDS and to lessen the stigmatization and fear associated with the condition.

In the previous chapter, I noted how local governmental practices are used to institutionalize the stigmatization of specific groups. For example, I discussed how the stigmatization of homelessness (i.e. viewing persons and places associated with homelessness as non-productive, dangerous, and personally culpable) led the City of Santa Ana to develop a local ordinance which constructed physical and legal barriers to the presence of homeless persons and places. One goal of this chapter is to provide an example of how intergovernmental practices at the federal, state, and local levels are working to promote acceptance and inclusion, rather than rejection and exile. Much of the literature investigating HIV/AIDS and community response has not focused on how acceptance and inclusion might be institutionalized through ongoing governmental policies. The potential for heightened acceptance through such strategies (which include community participation and localized data collection and analysis) seems too often overlooked as scholars study the possibilities of co-optation and of marginalization or oppression. A second goal of this chapter is to provide an understanding of the processes and outcomes of state response to HIV/AIDS which integrates institutional determinants with local context. I explore in this chapter how intergovernmental conflict and co-operation can result in inclusionary policies concerning persons affected by HIV/AIDS and those perceived to be at high risk.

One central component of efforts during the 1990s by the CDC centres on community participation and planning. Community participation and planning have long been argued to be a fundamental component of successful planning strategies. Since the 1960s, advocacy planning scholars have argued the necessity and centrality of community input and participation (e.g. Davidoff 1965; Peattie 1968; Silver 1985). The CDC requirement for community participation represents an innovation in that it has mandated nationally the incorporation of locally developed knowledge and the representation of affected groups in the development of public policy concerning HIV/AIDS education and prevention strategies. This policy shift is centrally relevant to understanding how stigma might be addressed, since it is at the local level, where context and institutional response become integrated through localized strategies for addressing HIV/AIDS, that public response to HIV/AIDS ultimately crystallizes and is expressed. I argue that contextualizing the institutional and political characteristics of governmental response with local experience provides insight in the mechanisms that bring stigma from theory to practice. Thus, in this chapter, I provide an overview of federal response and the experience from one state (California) and one municipality (Orange County, California) to illustrate where intergovernmental conflict and co-operation have led in terms of the sometimes competing goals of preventing HIV/AIDS and lessening its stigmatization.

Federal Response to HIV/AIDS: From Information to Decentralization

The first recognized case of AIDS was reported in the early 1980s. As public fear grew, public health efforts evolved in *ad hoc* and haphazard ways as community-based organizations and government agencies at the municipal, state, and federal levels struggled to respond to the widening crisis. Throughout the 1980s and 1990s, the federal government played a central role in developing a co-ordinated and comprehensive strategy for addressing the changing nature of HIV/AIDS. I focus in this chapter on the role of the Centers for Disease Control and Prevention (CDC), the leading federal agency for HIV/AIDS prevention, in shaping public efforts during the 1980s and early 1990s.

The CDC's earliest efforts centred on the development of epidemiologic studies and surveillance to track the spread of HIV/AIDS, and to identify the behaviours associated with high risk for transmission and infection (Noble *et al.* 1991). These studies indicated during the 1980s that there were significant public gaps in knowledge about HIV/AIDS, the modes of transmission, and effective prevention strategies. From the mid-1980s, the CDC responded with a variety of education and prevention programmes, with varying levels of success. By 1991, such programmes had evolved into three major types: (1) public awareness and information; (2) education for youth; and (3) education, counselling, and testing for individuals identified as being at high risk of contracting HIV (Roper 1991).

Public awareness and dissemination of information formed the basis of much of the early CDC efforts in education and prevention. There were various efforts during the late 1980s to disseminate information widely to the general public. Among the CDC's first HIV-related educational programmes was the National AIDS Hotline. Established in 1983, the Hotline was created to respond to the public's questions about HIV/AIDS, and to provide accurate AIDS-related publications and referrals to AIDS services. The National AIDS Clearinghouse was established in 1987 to catalogue and disseminate accurate information about HIV/AIDS (Sinnock *et al.* 1991). The first national public information campaign, 'America Responds to AIDS,' also began in 1987, and, in 1988, the national AIDS leaflet, 'Understanding AIDS,' was mailed to every household across the US. This government-produced pamphlet was aimed at the general public and contained essential facts about HIV/AIDS (Davis 1991).

Education and prevention of HIV/AIDS among youth also constituted a high priority for 1980s CDC programmes and funding. CDC's focus on education for school-aged populations began in 1986 when it first provided funding to state and local education agencies. In other efforts to target youth, CDC initiated co-operative agreements with national education organizations, colleges and universities, and local agencies to reach youth in high risk situations in 1987, 1990, and 1991. Risk reduction programmes constituted the third major component of CDC HIV/AIDS education and prevention policies during the 1980s and the early 1990s. Risk reduction education along with testing and counselling for at-risk populations was accomplished largely through co-operative agreements between CDC and state and local health agencies, and between CDC and community-based organizations.

During the 1980s and early 1990s, the organizational structure through which the CDC implemented HIV/AIDS programmes and funding revolved around four branches: National Center for Prevention Services (NCPS), National Center for Infectious Diseases (NCID), National AIDS Information and Education Program (NAIEP), and National Center for Chronic Disease Prevention and Health Promotion (NCCDPHP) (Table 9.1). Of these four branches, the National Center for Prevention Services branch (NCPS) was primarily responsible for the development of HIV/AIDS prevention programmes during the 1980s and early 1990s, focusing for the most part on technical and funding assistance to local level programmes in state and local health departments (Bailey 1991*a*).

The HIV/AIDS education and prevention programmes supported by CDC and the NCPS branch experienced a substantial increase in federal funding during the 1980s and 1990s. The HIV prevention budget at CDC grew from $200,000 in 1981 to $543 million in 1994 (Noble *et al.* 1991). Funding for HIV/AIDS prevention during the 1980s was made available to various grantees through the Public Health Services Act (PHS). NCPS's first national HIV/AIDS prevention programme was announced in 1985 and made funding available to states to establish alternative HIV testing sites (i.e. alternative to blood banks). Along with testing and counselling, NCPS developed programmes designed to provide HIV/AIDS-related education for populations identified as being at high risk of HIV infection. Findings from a 1984 CDC-funded study of local prevention efforts in nine US cities laid the groundwork for these more comprehensive education and prevention programmes (Bailey 1991*a*). This study argued for lifestyle appropriateness of prevention strategies tailored for specific groups, and for collaboration and co-operation among health agencies (federal, state, and local) and community-based organizations representing affected groups.

In 1985, NCPS announced the first programmes whereby state and local health agencies could enter into co-operative agreements with CDC. Eligible applicants for the 'Community Based Demonstration Projects' and 'Innovative Projects for AIDS Risk Reduction' were state and local health agencies, public or non-profit community-based organizations, educational institutions, or 'other organizations that [could] demonstrate the capacity of work in close cooperation with state and local health departments on the prevention and control of AIDS' (Bailey 1991*a*: 698). With these efforts, CDC hoped to encourage collaborative efforts and the exchange of information among organizations. The recipients of this initial and all subsequent funding through CDC were required to establish local review panels whose function was to approve the AIDS-related content of specific education and prevention programmes.

Health education and risk reduction co-operative agreements in 1986 between CDC and state and local health agencies expanded upon 1985

Public Policy

TABLE 9.1. *Organizational structure of Centres for Disease Control and Prevention*

Year initiated	CDC division	Programme description
1981	NCID	Epidemiologic studies
1982	NCID	AIDS case surveillance co-operative agreements with state and local health agencies
1986–7	NCID	HIV seroprevalence studies
1984	NCPS	Prevention co-operative agreements with US Conference of Mayors
1985	NCPS	Prevention co-operative agreements with state and local health agencies
1986	NCPS	Health education risk reduction
1986	NCPS	Counselling, testing, and partner notification
1987	NCPS	Minorities
1988	NCPS	Public information
1986	NCPS	AIDS community demonstration projects
1986	NCPS	Prevention co-operative agreements with Haemophilia Foundation
1986	NAIEP, NCPS	Contract supported National AIDS Hotline
1986	NCPS	Prevention co-operative agreements with state and local education agencies
1987	NAIEP	Contracted supported public information campaign
1987	NAIEP	Contracted supported National AIDS Clearinghouse
1987	NCCDPHP	Prevention co-operative agreements with national education organizations
1988	NCPS, NAIEP, NCCDPHP	Prevention co-operative agreements with national and regional minority organizations
1988	NAIEP	Prevention co-operative agreement with American Red Cross
1988	NCPS	Prevention co-operative agreements with community-based organizations
1989	NAIEP	Prevention co-operative agreements with national organizations
1989	NCPS	Comprehensive community based HIV programme
1990	NCCDPHP	Prevention co-operative agreements with colleges and universities to reach college age youth
1991	NCCDPHP	Prevention co-operative agreements with local agencies to address youth in high-risk situations
1991	NCCDPHP, NCPS	Prevention of HIV in women and infants demonstration projects
1991	NCPS	Co-operative agreements with state and local health agencies, TB demonstration

CDC programme components: NCID—National Center for Infectious Diseases, NCPS—National Center for Prevention Services, NAIEP—National AIDS Information and Education Program, and NCCDPHP—National Center for Chronic Disease Prevention and Health Promotion programme efforts, provided funding for state-based HIV/AIDS projects, and continued funding for education and prevention and programme evaluation in geographic areas with a high incidence of AIDS. In 1987, NCPS attempted to spur state and local health departments to respond to the changing epidemiology of HIV/AIDS by providing funding to address specifically the education and prevention needs of populations of colour (especially African Americans and Hispanics), school-aged youth, and health providers.

But dissatisfaction among service providers and municipal health care agencies began to mount. States were chosen to receive and further disseminate funds for the newest programmes targeted at persons of colour, causing dissension at the local level because the states had been selected to administer this funding (rather than either directly disseminating funds to organizations or choosing more local administrators). In addition, the *ad hoc* and haphazard development of federal policies meant that the bureaucracy charged with administering and monitoring funds was increasingly onerous. Representatives of affected groups, community-based organizations, and locally elected officials had become frustrated by the administratively burdensome Request for Proposal (RFP) process and the complex funding requirements associated with receiving funding from the states; they began to call for more localized control of funding. In response to these demands and widening dissatisfaction, CDC began a programme of direct funding to community-based organizations in 1988 (Bailey 1991*b*).

This history of CDC's role in HIV/AIDS education and prevention indicates that the federal agency was dynamic in many ways in responding to the changing epidemiology of HIV/AIDS and the frustration of health departments and community-based organizations. However, the partnerships between CDC and state and local health departments that emerged out of CDC's 1980s and early 1990s efforts to curb transmission of HIV/AIDS remained problematic in three ways. First, the proliferation of CDC initiatives had created multiple funding streams. For example, in 1991–2, funding from CDC to states involved several different CDC branches, prevention programmes, and funding routes. There were funding streams from: (1) CDC to states via co-operative agreements; (2) CDC to cities via co-operative agreements; (3) CDC to community-based organizations; (4) states to cities; (5) states to community-based organizations; and (6) cities to community-based organizations. The multiple CDC initiatives and corresponding funding routes led to confusion and competition rather than the co-operation that CDC had originally intended. In this climate, prevention programmes were often implemented without co-ordination, and competition for funding meant there was little incentive to consolidate existing programmes to develop effective prevention strategies. Second, CDC requirements that prevention funding be allocated for surveillance, counselling, testing, and partner notification meant that the majority of CDC funds were not being used for comprehensive, community-based HIV/AIDS prevention strategies (Institute of Health Policy Studies 1993). Third, local review panels (required by CDC in its funding guidelines) directly intervened in the development of targeted, explicit HIV/ AIDS prevention messages. Critics of the local review panel structure believed that such panels were causing the homogenization of information and education 'to meet the test of being tolerable to a majority rather than the educationally superior approach of materials designed to appeal to a specific target audience' (US House of Representatives 1992).

In response to these three problem areas, and following the precedent set by the Ryan White CARE Act[1] which instituted community planning for the provision of AIDS-related services, CDC instituted a new programme in December 1993. Although CDC had always intended as a cornerstone of its HIV/AIDS prevention policy the collaboration among different levels of government, community-based organizations, and representatives of affected populations, these groups were not compelled or coerced into co-operative agreements or co-ordinated efforts until CDC introduced this new programme. In its new programme, to encourage co-operation and collaboration, CDC instituted a community planning process, restructuring existing federally funded HIV/AIDS prevention programmes by requiring representative community input and the use of locally developed data and scientific evidence in decision-making (Valdiserri *et al.* 1995). In developing this new approach, CDC sought to provide flexibility in two ways. First, the programme allowed for local priorities to take precedence in determining how prevention funds should be spent by eliminating the requirements that a certain percentage of federal funding be allocated for testing and counselling. Second, rather than mandate one specific approach to HIV/AIDS prevention planning, this programme issued thirteen principles to guide the planning process:

1. Community planning represents on ongoing process—not a one-time event.
2. Differences in background, perspective, and experience are essential and valued in the planning process.
3. Planning is characterized by shared priority-setting.
4. Each grantee must have at least one specifically designated prevention planning group which reflects in its composition the characteristics of the epidemic in that jurisdiction.
5. Nominations for membership are identified through an open process.
6. Roles and responsibility are clarified at the outset.
7. Accurate epidemiological assessment is the starting-point for defining HIV-prevention needs.

8. Identification, interpretation, and prioritization of HIV-prevention needs must reflect culturally relevant information.
9. Needs assessment is based on a variety of information sources and collection strategies.
10. Priority-setting for specific prevention strategies and interventions is based on documented need, science, consumer preference, and local circumstances.
11. Resources are used to facilitate the involvement of community representatives in the planning process.
12. Policies and procedures for resolving disputes and avoiding conflict should be developed proactively.
13. Evaluation is an essential component of the HIV prevention community planning process. (Valdiserri *et al.* 1995)

But the implementation of these principles and local-level participation has been widely variable among and within states. I focus on the experience in the State of California and the municipality of Orange County, two jurisdictions which have been recognized as leaders in HIV/AIDS education and prevention. I turn first to the institutional context of the California State bureaucracy, then discuss the recent experience of Orange County in developing an HIV Prevention Plan which was completed in December 1995.

A Proactive State: The Institutional Context of HIV/AIDS Prevention in California

HIV/AIDS has been a critical social and health issue across the State of California, but has been particularly acute for its large metropolitan areas. Prior to 1983, there were fewer than 300 AIDS cases statewide. At the end of 1990, the number of cumulative reported AIDS cases stood at nearly 38,000, with the number growing to nearly 100,000 during 1996. Reported AIDS cases have tended to be concentrated in two geographic areas, San Francisco and Los Angeles, with Whites and males the most impacted groups. However, mirroring national epidemiological statistics, African Americans and Hispanics have also been significantly impacted by HIV/AIDS in California. Reported AIDS cases throughout the state have tended to be concentrated in Whites (64 per cent), Hispanics (18 per cent), and African Americans (16 per cent) (Department of Health Services 1996*b*).

The California Department of Health Services, State Office of AIDS (henceforth State Office of AIDS) oversees HIV/AIDS policies in the state (Fig. 9.1). Three branches comprise the State Office of AIDS: Education and Prevention Services, Epidemiology Services, and Pilot Care and

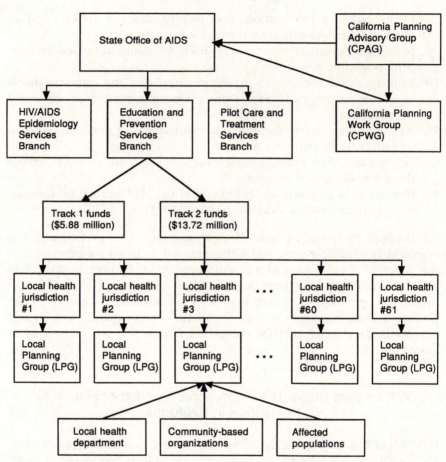

FIG 9.1. California State Office of AIDS organizational structure

Treatment Services. Among its many functions, the State Office of AIDS implements the California AIDS Clearinghouse which serves as a resource centre for state contractors and health care providers. The State Office of AIDS also works closely with its 61 local health departments, comprised of 58 counties and 3 localities (Pasadena, Long Beach, and Berkeley), to assist in providing and co-ordinating HIV/AIDS-related services and prevention efforts at the local level. Although all 3 of the branches of the State Office of AIDS deal with HIV/AIDS prevention, I will focus in this chapter on the Education and Prevention Services branch, which has felt the greatest impact of the CDC's HIV Prevention Community Planning programme.

The Education and Prevention Services branch administers federal and state funds available to the State Office of AIDS for HIV/AIDS education and prevention. In 1994, $30.4 million was apportioned among four divisions within this branch: HIV/AIDS education and prevention programmes ($16.7 million), HIV testing and counselling ($10.0 million), training programmes ($0.5 million), and early intervention strategies ($3.2 million). Education and prevention funds are not limited to those received through this branch, however. There is also a substantial amount of funding earmarked for education and prevention services received by the state that is not under the jurisdiction of the State Office of AIDS, Education and Prevention Services branch. In 1994, for example, the state received an additional $36.2 million, mostly from federal sources, for HIV/AIDS education and prevention in California. These funds included direct federal funding to community-based organizations ($10 million), CDC co-operative agreements to Los Angeles and San Francisco ($13.4 million), and funding for HIV testing under the auspices of the federal agency SAMHSA (the Substance Abuse Mental Health Services Administration).[2]

Of the four divisions in the Education and Prevention Services branch, the Education and Prevention Program received over 50 per cent of the funding in 1994.[3] This funding was earmarked for use by local public and private organizations in the development and implementation of HIV/ AIDS prevention efforts. Prior to the implementation of HIV Prevention Community Planning in California (mandated by the CDC), the Education and Prevention Program funded 150 projects including 38 contractual agreements with health departments and 70 community-based organizations. Twenty-three of the 38 contractual agreements with local health departments involved collaborative agreements with community-based organizations where the health departments acted as lead agencies within the jurisdictions, subcontracting with the community-based organizations to provide education and prevention services to specific target populations. To receive funding, grantees were required to submit a competitive proposal through an Invitation for Bid (IFB) process (the State of California's version of a Request for Proposal). But, as I previously mentioned, the use of the IFB process had become highly contentious prior to the initiation of the process of community planning because of the inherent competition involved, the rapidly changing nature of the epidemiology of HIV/AIDS in the state, and the number of groups responding to the IFB (California Community Planning Working Group 1995).

The State Office of AIDS began to implement the HIV Prevention Community Planning Process in December 1993. Two groups were formed in California under the auspices of the Community Planning Process (Fig. 9.1). The first, HIV Prevention Community Planning Advisory Group (CPAG), served as an advisory board to the State Office of AIDS; it included twenty members whose input helped to determine the selection of the second group, the Community Planning Work Group

(CPWG). The CPWG was comprised of fifty-three voting members who represented community-based organizations, advocacy groups, state and local government agencies, university researchers, legislative staff, national and state AIDS organizations, and HIV/AIDS-affected communities. This group, following CDC guidelines, produced the 1995 California HIV Prevention Plan. The Plan was extensive in scope and content to encompass the goal of providing a framework for funding and developing future prevention programmes, with support from both federal and state prevention funds.

The CPWG, although not specifically required to by CDC guidelines, developed recommendations for funding allocations and processes in its HIV Prevention Plan for California (California Community Planning Working Group 1995). The CPWG reviewed the entire $35.2 million available to the State Office of AIDS, Education and Prevention Services branch for prevention funding during the fiscal year 1995–6; it recommended that funds be allocated among the four programmes in the following manner: $19.6 million (56 per cent) to education and prevention; $10.6 million (30 per cent) to HIV testing; $0.5 million (2 per cent) to training; and $4.5 million (13 per cent) to early intervention. The CPWG also made recommendations regarding the methods used to distribute funds within the four programmes. Funding processes were recommended to be maintained for HIV testing and training programmes, and to be altered for the education and prevention and early intervention programmes.

The CPWG suggested that Early Intervention Program funds be distributed using a competitive Invitation for Bid (IFB) process with proposals reviewed by a 'culturally competent' panel (i.e. a panel familiar with the specific language, cultural, and traditional practices of specific target groups), and that the IFB process for education and prevention funding be substantially changed to address concerns which had been raised by health care agencies and community-based service providers. As previously discussed, the use of the IFB process to award funding for local education and prevention initiatives had become contentious. Projects based on this old funding allocation system were to end on 30 June 1995, and funding for new projects was to be dictated by the statewide prevention plan. To address the transition in funding allocation processes and the contentiousness associated with the prior funding allocation mechanism, the CPWG recommended that Education and Prevention Program funds for the fiscal year 1995–6 be distributed in two tracks. Track 1 funds ($5.88 million, 30 per cent) were to be allocated for statewide mandated programmes and statewide/regional set asides via an Invitation for Bid (IFB) process.

Track 2 funds ($13.72 million, 70 per cent) were to be awarded to local planning groups (LPGs) according to a weighted statewide formula. The formula was comprised of the following elements and weights: incidence of AIDS cases (40 per cent), percentage of population from communities of colour (25 per cent), percentage of population below poverty level (15 per cent), percentage of population living in rural areas (15 per cent), and percentage of neonatal surveillance (5 per cent). Track 2 funds required the establishment of local planning groups that would decide locally how funds should be used and distributed. Under this Track, the local health department or its designee in each of the 61 local health jurisdictions (58 counties and 3 localities) in California would be the fiscal agent for the community planning process. According to the State of California HIV Prevention Plan, a local advisory committee would select members of the LPG, and decision-making powers and responsibilities would then be shared between the fiscal agent and the LPG. In each of the 61 local health jurisdictions, the entire planning process was to be modelled on the state planning process and CDC guidelines.

Opportunities and Challenges of Community Planning

Community planning has a high potential for success across multiple dimensions in local HIV/AIDS policies. In terms of HIV/AIDS prevention, community planning, because of its targeted strategies and locally developed programmes, has the potential for effectively reducing HIV transmission. There are also broader implications of this process for inclusion and political participation among those affected by HIV/AIDS. To the extent that community planning is inclusive, there is the potential for groups such as persons living with HIV/AIDS, groups identified as being high risk, and other concerned individuals and groups to affect materially the substance of HIV/AIDS prevention messages (thereby altering the representation of HIV/AIDS) and further to influence the funding streams associated with federal funding mechanisms. In addition, there are institutional implications, as the community planning process works for co-ordination and co-operation among health departments, community-based organizations, and affected individuals and groups. Collaboration among these groups in providing HIV/AIDS prevention services means that there is the potential for inclusion of persons affected by HIV/AIDS (thereby potentially altering the politics of HIV/AIDS at least at the local level), consolidation of competing and overlapping programmes, and greater cost effectiveness within this locally based process. But while the community planning process offers many potential benefits, there are also substantial challenges to its continued development, including the difficulties in promoting co-operation, the conflict between governmental and non-governmental perspectives, and the achieving of adequate and appropriate representation from the community.

One of the primary challenges continued to be faced in the community planning process is the difficulty in creating a sense of community identity and cohesiveness which supersedes individual group identity, that is, the promotion of collaboration among groups rather than competition. The diminishing availability of funding support and the past and current institutional practices underlying federal funding have contributed to this ongoing climate of competition. The shrinking of available public and foundation funding for community-based organizations and local agencies for human service provision over the past decade has meant that groups are increasingly in direct competition for financial support (Wolch and Dear 1993). In addition, the institutional practices of federal funding have created a context where competition among local groups is the norm. Representatives of distinct groups, accustomed to working in an atmosphere where ardent advocacy constitutes the best means of securing prevention funding for their groups, may find difficulty in putting aside an advocate identity in favour of a community-wide focus.

Another major challenge is the inherent difficulty of a participatory planning process which involves both governmental and non-governmental perspectives. For example, government officials may not perceive community participants as having a legitimate voice in technical issues (Valdiserri *et al.* 1995). This challenge may be significant in processes where technically complex issues such as priority-setting constitute the primary goal. Individuals not previously exposed to these types of planning and prioritization procedures may find the process formidable and incomprehensible, especially if government actors use such technical complexity as the means of excluding community-based and HIV-impacted representatives.

The final central challenge to the process of community planning is the selection of actual planning group participants who represent the community. The HIV/AIDS prevention environment (and the political dimensions of HIV/AIDS more generally) is extremely complex and dynamic, with numerous actual and potential participants and special interests. With the changing and dynamic dimensions of HIV/AIDS, the identification and solicitation of persons with significant and relevant viewpoints may be problematic (West and Valdiserri 1994). In addition, members of groups traditionally marginalized by mainstream society may be reluctant to participate in the planning process because of fears of public knowledge regarding their personal situations, specifically their HIV status and the social implications HIV-positive status brings (Valdiserri *et al.* 1995; Wilton 1996).

The effectiveness of these alternative viewpoints on HIV/AIDS prevention strategies may in addition be limited by prevailing public opinion or because of the institutional inertia which continues to exist in terms of funding and service provision (Bennett and Sharpe 1996; Cain 1995). The representation of HIV/AIDS as White gay men may play a distinct role in the determination of alternative viewpoints. If alternative viewpoints are defined as varying from the perspectives of the medical community, government actors, and service providers, then particular perspectives, namely White gay men, may appear to be the appropriate choice to represent the views of the HIV/AIDS affected population. The social construction of HIV/AIDS as White gay men, while incorporating (although perhaps not necessarily empowering) one perspective, that based on the dimension of HIV/AIDS as homosexuality, may actually serve to institutionalize further the marginalization of others (such as persons of colour and women). Thus, the political mobilization of one marginalized group, White gay men, largely through the institutionalized identity politics of HIV/AIDS, may lead to further marginalization of various Other groups, including but not limited to gay men of colour, women living with HIV/AIDS, and women of colour living with HIV/AIDS. Even with these limitations, however, the community planning process represents the first time in HIV/AIDS education and prevention that community participation has been mandated by a federal agency for funding, and thus represents a unique opportunity to assess the opportunities and challenges offered by federal, state, and local efforts to engage the stigmatization of HIV/AIDS.

Community Planning in Orange County, California

HIV/AIDS in Orange County, California

Orange County, California (widely known as the home of Disneyland), may appear an unlikely location to study the impact of HIV/AIDS, and the role of community participation in its prevention. However, like its more urbanized neighbour, Los Angeles, Orange County has experienced an increasing incidence of HIV/AIDS among its populace. By 1996, there had been nearly 4,400 reported AIDS cases throughout the County. Among Orange County males aged 25–44 years, AIDS was the leading cause of death in 1991 and 1992, and the second leading cause of death in 1993. The significance of HIV/AIDS for the County is also exhibited in the cost of health care. Charges for AIDS-related admissions at Orange County Acute Care Hospitals totalled nearly $29 million in 1992 (Orange County HIV Prevention Planning Committee 1995).

Orange County is far from the monolithic suburban, theme park economy conjured up as an image by many living outside its borders. The County has experienced massive population growth and demographic change over the past two decades (Baldassare and Wilson 1995). Now the fifth most populous county in the US with over 2.4 million persons, Orange County added nearly half a million individuals between 1970 and 1980, and grew at a rate of 25 per cent (about 475,000 new residents) between

1980 and 1990. The 1980s were also years of increasing racial and ethnic diversity. From 1980 to 1990, the percentage of Hispanics, Asians, or African Americans grew from 21 per cent to 35 per cent, with most of the growth in the Hispanic and Asian populations the result of immigration. By 1990, Whites made up 65 per cent of the County's population, Hispanics 23 per cent, Asians 10 per cent, and African Americans 2 per cent.

Well known for its conservative politics and Republican voting patterns (Baldassare 1990), Orange County has received a great deal of attention during the past few years due to its $1.69 billion investment fund loss and subsequent bankruptcy. The bankruptcy has fuelled concerns among County residents and community-based organizations alike concerning public service provision. Among residents were concerns about the negative repercussions on County services and schools (Lee 1995). Among community-based organizations, especially those involved in human services provision, were widespread concerns that they would be paid in a timely fashion if at all (Orange County HIV Planning Advisory Council 1996).

In the midst of rapid population growth and change, and municipal financial distress, the County has also experienced a growing and diversified population affected by HIV/AIDS. The epidemiological portrait of AIDS indicates that AIDS tends to be concentrated in Whites and Hispanics; Whites comprised 72 per cent of reported AIDS cases in Orange County compared to 47 per cent nationwide (Table 9.1). When compared to reported AIDS cases throughout California, Orange County's reported AIDS cases have greater proportions of Whites (72 per cent) and Hispanics (21 per cent) and far fewer African Americans (4 per cent), reflecting in part the demographic composition of the County. Similarly to the state, nearly 80 per cent of persons living with AIDS in Orange County contracted HIV/AIDS via homosexual or bisexual contact, in sharp contrast to the national figure of 51 per cent. In addition, the 11 per cent of cases due to exposure via injection drug use in Orange County is much closer to the California figure of 10 per cent than to the nationwide figure of 25 per cent. Thus, the epidemiology of AIDS in Orange County, both the demographic profile of those affected and the modes of transmission, are very similar to California's portrait and very different from the nation as a whole.

The Community Planning Process

With the CDC's requirement for community planning for HIV programme funds, the State of California passed responsibility for education and prevention to its sixty-one local health jurisdictions, clearing the way for the determination of HIV/AIDS prevention needs at the local level. The state in its guidelines required that a local advisory committee select members of the local planning group (LPG), and that existing AIDS-related planning bodies serve as the advisory committee if possible (Department of Health Services 1995). Orange County adhered to this recommendation in developing the institutional structure for its community planning process. The existing Orange County HIV Planning Advisory Council (henceforth the Advisory Council) assumed responsibility for developing the Orange County HIV Prevention Plan, becoming the local planning group (LPG). The Advisory Council has forty members, their characteristics closely mirroring the demographic profile of the County's reported AIDS cases.

Since assuming responsibility for HIV/AIDS community prevention planning and being named the local planning group (LPG), the Council has also retained three other major activities and responsibilities: (1) as Ryan White Title I HIV Health Services Planning Council; (2) as Ryan White Title II Consortium; and (3) as advisory committee to the County Health Officer. In addition, the Advisory Council establishes priorities for the expenditure of Housing Opportunities for People With AIDS (HOPWA) funds that are administered by the City of Santa Ana (the County seat).[4] To maintain timely implementation of these multiple roles and responsibilities, the Advisory Council instituted an annual HIV Services Planning Process, resulting in a single list of service priorities and an implementation plan for Ryan White I, II, and HOPWA funds (Orange County HIV Planning Advisory Council 1996).

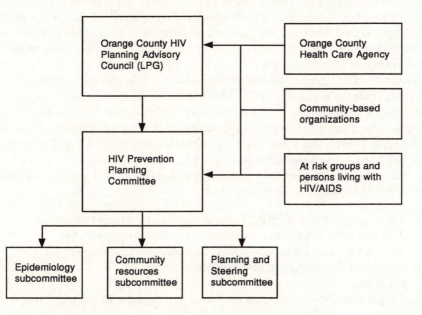

FIG 9.2. Community planning process, Orange County, California

Because of its multiple responsibilities, the Advisory Council delegated the community planning process to its HIV Prevention Planning Committee (Fig. 9.2). The first step in the planning process was the formalization of HIV Prevention Planning Committee membership. One-third of the final committee comprised governmental representatives, and two-thirds comprised non-governmental, community representatives. Membership of this new HIV Prevention Planning Committee which would be responsible for developing the Plan was developed by a membership subcommittee of the existing HIV Prevention Planning Committee. The subcommittee reviewed State Office of AIDS guidelines and made membership recommendations to the existing HIV Prevention Planning Committee. In addition to existing members, membership now also included representatives from the community (individuals representing the socio-demographic profile of the local population living with HIV/AIDS), additional service providers, and behavioural scientists/academics. The Orange County HIV Prevention Planning Committee and three subcommittees (Planning/Steering, Community Resources, and Epidemiology) began meeting in the spring of 1995 with the intention of producing a plan by December of that same year, providing a comprehensive framework for preventing the transmission of HIV/AIDS in Orange County (Orange County HIV Prevention Planning Committee 1995).

The HIV Prevention Planning Committee successfully created a comprehensive plan within the relatively short time frame required by the state. The Epidemiology subcommittee prepared an extensive epidemiological and demographic portrait of Orange County. This portrait was to serve as the scientific foundation upon which decisions could be made regarding the prioritization of target populations and the allocation of resources. The Community resources subcommittee developed a catalogue of the County's existing HIV/AIDS education and prevention service providers, their programmes and funding, and targeted and served populations. The Planning/Steering subcommittee developed strategic plans for the HIV Prevention Planning Committee's administrative and organizational policies, and provided linkages with state committees.

The community planning process resulted in several policy outcomes which will be used to distribute funds:

1. Final membership of HIV Prevention Planning Committee:
   33.3 per cent of members were governmental representatives
   66.7 per cent of members were community representatives
2. Extensive demographic and epidemiological profile of Orange County prepared.
3. Target populations selected and prioritized in the following order: men

who have sex with men, persons of colour, substance users, youth, women, and others (heterosexuals and homeless persons).
4. Needs assessment of target populations conducted.
   Information beyond the epidemiological profile collected from the following sources: (*a*) interviews with HIV prevention providers; (*b*) focus groups with specific sub populations; (*c*) Latino Community Forum.
5. Specific HIV prevention strategies and interventions not prioritized, but criteria to be used in selected interventions for identified target populations developed.
6. Linkages between primary and secondary interventions assessed.
7. Goals and objectives developed and approved:
   Goal 1. Decrease HIV infection in Orange County;
   Goal 2. Enhance capacity of Orange County prevention providers to develop quality programmes and services.
8. Issues of co-ordination and linkages assessed and the following plan to maintain a co-ordinated prevention strategy developed:
   A. Maintain current committee membership;
   B. Extend membership to include other groups (suggestions made);
   C. Prevention Planning Committee in Orange County should have one member who also serves on the California CPWG to insure linkage between state and county.
9. Top technical assistance needs identified and plans for obtaining relevant technical assistance suggested. Top technical needs identified were:
   A. Staff training;
   B. Developing and obtaining HIV materials;
   C. Building community coalitions;
   D. Development of effective programmes.
10. Resource allocation process developed with percentage of resources to be allocated among target populations specified (see no. 3 above) and decision to use Request for Proposal (RFP) to award prevention funds made.
11. Planning process evaluated.

Several of the most important outcomes of the community planning process concerned the nature of the specific target populations which would receive HIV/AIDS education and prevention services, and the appropriate strategies used to provide these services. After a needs assessment was conducted by an outside consulting agency, and after an extensive review of local epidemiologic trends and existing and needed community resources, the HIV Prevention Planning Committee in its entirety selected target populations and prioritized them. An outside consulting organization facilitated the technical process used to select target populations to receive HIV education and prevention services in the County. Target groups were prioritized in the following order: men who have sex with men, persons of colour, substance users, youth, women, and others (including heterosexuals and homeless persons). The needs assessment resulted in information used by the HIV Prevention Planning Committee to develop criteria for evaluating prevention strategies and interventions. Following these determinations, the HIV Prevention Planning Committee collectively decided upon ways that education and prevention resources would be allocated.

Outcomes of the community planning process were both philosophical and pragmatic: the establishment of goals and objectives for HIV/AIDS education and prevention for the County, the identification of technical assistance needs, and the assessment of linkages between primary and secondary interventions.[5] There were also organizational reviews and recommendations. To facilitate achieving state and local HIV/AIDS prevention goals, for example, co-ordination and communication linkages were assessed and suggestions were made to maintain a co-ordinated prevention strategy.

The overall plan developed by the Orange County HIV Prevention Planning Committee was rated as 'excellent' by a team of state reviewers (Department of Health Services 1996a). However, this community planning process did not transpire without problems. An evaluation of the process underlying the development of the plan was conducted by an outside consultant and identified several challenges (which were identified in the County's final plan along with suggestions to address these issues). The State Office of AIDS's external evaluation of the plan, in most instances, simply recommended that the suggestions for improvement already outlined in the plan be implemented.

Many of the problems in the community planning process concerned the nature of power relations within the HIV Prevention Planning Committee (see also Schietinger *et al.* 1995). Specifically identified were the selection of members for the HIV Prevention Planning Committee, the perception of unequal influence by individual members on committee decisions (particularly the perceived greater influence by members representing the County), and access to technical information and knowledge concerning HIV/AIDS. The short time frame mandated by the state for producing the local prevention plan meant that an open nomination process was not feasible to select members of the HIV Prevention Planning Committee. Power relations within the committee were also made problematic as many HIV Prevention Planning Committee members often personalized their roles and remained intrinsically tied to the populations they represented. Some members attributed this lack of community identity to the lack of development of clear roles and responsibilities during the initial planning phase. The inertia of prior institutional funding practices and the lack of creation of new roles and responsibilities coinciding with this new community planning process meant that members at times acted in their familiar advocate roles rather than as members of a collective.

The shadow of competition for funding, and the potential suspicion initiated by the membership of the committee (i.e. that the County's influence could control committee decisions), meant that discussions about target populations and funding brought about contention and conflict which had been relatively absent in the initial stages of the community planning process (e.g. during the process when the epidemiological profile and community resources survey were being developed). Many committee members had a difficult time putting aside the notion that prioritization of target populations was going to determine how education and prevention funds would be allocated. The perception of unequal power relations was evident in criticisms by several HIV Prevention Planning Committee members concerning the influence of the Orange County Health Care Agency (the fiscal agent in the process). Several members argued that although members on the HIV Prevention Planning Committee had equal votes, they did not have equal power in decision-making (i.e. that the members of the HIV Prevention Planning Committee representing the Orange County Health Care Agency had substantially more influence than other individuals and community-based organizations).

Such issues resurfaced eighteen months after completion of the Plan. In subsequent HIV Prevention Planning Committees during 1997, as the fiscal agent (Orange County Health Care Agency) worked to disseminate funds and monitor programme effectiveness, there arose questions about where disputes about target populations and programme methods should be addressed. For example, the environment of decreasing federal funding made HIV/AIDS education and prevention even more competitive, and local service organizations looked to ways to expand the populations they served. This expansion of target populations by organizations meant an increasing potential for overlap among existing service agencies as they tried to adapt to the changing nature of governmental and foundational funding priorities. Although the funding priorities established in the Plan through the community planning process remain the same, their level of vagueness (e.g. identifying a target population as 'persons of colour') has meant that implementation is very dependent on the interpretation by the County, by the LPG, or by the HIV Prevention Planning Committee.

Such ensuing conflict has also been influenced by broader federal funding changes. Federal agencies continue to adapt and change their funding streams, requiring that Orange County and community-based organizations adapt their fundraising strategies to meet such changing requirements. For example, the CDC in early 1997, after mandating community-based planning for HIV education and prevention in 1993 which had been a response to the problems developing since the mid-

1980s, released a new Request for Proposals which would provide funding directly to collaborative community-based efforts for education and prevention programmes for populations of colour. Thus, community organizations, especially those concerned (or becoming involved) in education and prevention services for persons of colour, have had to remain flexible and adaptable to changing federal funding streams, while at the same time responding to state and county level requirements for existing programmes outlined in the County's Prevention Plan.

Conclusion

Efforts by the Centers for Disease Control and Prevention (CDC), implemented through the State of California, to institutionalize local knowledge and community participation in the development and implementation of HIV/AIDS prevention accomplished multiple goals in Orange County, California. First, there were substantive data collection results. Local epidemiological data were collected and analysed to form the basis for developing a prioritization of target groups for education and prevention services, and a catalogue of community-based organizations and their services. Second, there were representational and participatory results. Most importantly, affected groups (e.g. persons living with HIV/AIDS and groups identified as being at high risk of HIV transmission), who had had minimal presence in previous decision-making processes concerning the development, funding, and implementation of education and prevention policies, were directly involved in the formulation of the Prevention Plan which the County now uses to distribute funds. Third, institutional structures were created at the local and state levels. At the local level in Orange County, the HIV Prevention Planning Committee (following the guidance of the Advisory Council acting as the local planning group) continues to develop and adapt its planning process (in response to changing CDC and state guidelines) to encourage greater community participation and appropriate representation, to develop stronger linkages with state committees and agencies, and to ensure that adequate data and widely accepted methods are used to prioritize needs and evaluate results. At the state level, fifty-four local planning groups (LPGs) were established, creating a network of local groups linked with state organizations to develop prioritized needs and to disseminate state funds. The State Office of AIDS continues to provide external reviews of community planning processes, and to provide technical assistance and information for LPGs to facilitate community planning into the fiscal year 1996–7 and beyond.

There were several elements which facilitated this community planning process in Orange County with relatively minimal conflict. First, the short time frame allowed for the development of Orange County's HIV Prevention Plan necessitated the putting aside of many significant potential and existing suspicions and conflicts because of the pragmatic and collective need to accommodate the state's and CDC's guidelines for data collection, analysis, and evaluation. Because funding for all organizations and individuals was tied to the completion of this plan, members of the HIV Prevention Planning Committee for the most part engaged in minimal partisanship in favour of completing the plan on schedule.

The spring to December time frame also functioned to minimize debate over membership on the HIV Prevention Planning Committee. The process of selecting members may also have served to minimize conflict during the development of the plan, since organizations and individuals not already involved in local education and prevention, or those deemed to be controversial, confrontational, or non-cooperative, did not gain entry to the decision-making process. The issue of membership was revisited by the Orange County HIV Prevention Planning Committee during the spring of 1997 as renewed discussion and debate occurred over the composition and privileges of members.

Second, the relatively few high risk groups and the relatively small group of HIV/AIDS education and prevention providers in Orange County meant that there were fewer issues of contention than for larger more diverse health jurisdictions. Although there was conflict during the identification of prioritized target groups and the determination of funding criteria, there would have been much more given more groups in need and an expansive network of service providers. Moreover, since Orange County used an existing planning body for the development of the plan, an already recognized institution existed for the determination of target groups and funding criteria. The legitimacy of this planning body reduced the potential conflict among services, and enabled co-ordination among community-based organizations, County staff, and representatives of affected groups.

Third, the dynamic nature of HIV/AIDS (characterized through local epidemiology and local service needs) has meant that this planning process should retain as much flexibility as possible to fulfill ongoing concerns about cost effectiveness and targeted service provision. Thus, there has tended to be little inertia concerning the substance and types of education and prevention services, as service providers and the County have sought more effective means of reaching their target populations and providing them with education and prevention services. This lack of programmatic inertia and ongoing need for education and prevention created an atmosphere where HIV Prevention Planning Committee members were collectively searching for more effective means of identifying populations in need, of contacting these populations, and of creating education and prevention strategies appropriate to the diverse language, cultural, and traditional needs of varying groups. However, as specific target groups were identified through the planning process, available funds have been prioritized for such programmes, making organizations begin to shift their programme emphases. While this may have benefits concerning meeting service gaps throughout the County, this linking of target groups and funding has also meant that organizations are expanding their current target populations, spawning a potentially growing conflict among service providers about which organization(s) are the best (and therefore should receive funding) for providing services to men who have sex with men, persons of colour, and the other targeted populations.

However, in general, community participation in the Orange County HIV Prevention Community Planning process, although still problematic, created a relatively collaborative atmosphere for individuals representing affected groups, community-based organizations, and Orange County Health Care Agency staff. The ongoing institutionalization of this community participation for CDC and state funding for HIV/AIDS education and prevention programmes will probably result in changing relationships and membership within the HIV Prevention Planning Committee, and, perhaps, a greater bureaucratization of community planning. This makes central the question of the material effectiveness of community participation in this process; does community participation actually affect the nature of service provision, and, more broadly, the stigmatization of HIV/AIDS, or does community participation serve to legitimate existing institutional practices?

In this process of community planning, participation appeared to serve both functions. The participation by individuals and groups in the development of the Orange County plan who had had minimal influence on previous education and prevention funding served to focus attention on new and emerging target populations, for example, in Orange County, Latinos/Latinas and young people. Debate and discussion, often vigorous at times, served to enlighten HIV Prevention Planning Committee members about the distinct perspectives represented by these groups; this contributed to a greater inclusiveness of views and potential redistribution in funding resources, along with the fragmentation of representation concerning HIV/AIDS (i.e. from the widespread understanding of HIV/AIDS as White gay men, even among many service providers, to encompass a more diverse population by race and gender).

Participation, however, also served to legitimate existing institutional structures and practices. Critical comments by several HIV Prevention Planning Committee members highlighted the significant influence of the staff of the Orange County Health Care Agency on the planning process, through membership composition, existing programme size and resources, and technical expertise and access to information. Thus, although there was representation by multiple groups and perspectives, the contribution of these perspectives must be tempered by the pragmatic issue of membership and voting privileges, and the nature of existing organizational structures linking the Orange County Health Care Agency to a network of community-based organizations.

The community planning process outlined by the CDC, the State of California, and the Orange County Health Care Agency itself, has made possible internal and external critiques and adaptation of the process. Thus, although there remain concerns about the nature of participation and representation, the institutionalization of community participation in HIV/AIDS education and prevention holds much potential in not only improving the cost effectiveness of education and prevention programmes, but also allowing local knowledge and representation eventually to influence the ways in which the state and the CDC design, fund, and implement HIV/AIDS policies. Such efforts are a central element not only in shifting public policies towards local needs, but also in developing a localized representation of HIV/AIDS.

Part V
Conclusions

10

Problematizing Fairness

FAIRNESS in the context of the NIMBY syndrome has become increasingly problematic as diverse groups have immersed themselves in community opposition over human service facility siting. Historically disadvantaged communities have increasingly entered the fray of locational conflict over the siting of facilities associated with homelessness and HIV/AIDS. But to complicate matters even more, home-owning and suburban residents too have claimed the NIMBY syndrome as expressive of grass-roots attempts to engage in the democratic process. That is, both disadvantaged and privileged communities have begun to use the language of social movements, community mobilization, and oppression to bolster their arguments concerning the undesirability of specific facilities. Although the NIMBY syndrome remains widely understood as consisting of privileged groups defending, maintaining, and reinforcing the existing system of social relations, both privileged and disadvantaged communities have begun to (re)define NIMBY strategies as collective social mobilization, arguing that such efforts defend and expand local autonomy from institutions and structures deemed to be marginalizing and oppressive. Thus, even communities of privilege have turned to the language of 'facility siting as marginalization' to defend their actions, promoting NIMBY as populist democracy, and, further, as practising the constitutional rights of free speech and dissent.

The case of three residents in Berkeley, California, opposing the proposed conversion of a motel into housing for homeless persons provides a stark illustration. This dispute, publicized in the national publications, *U.S. News & World Report* and *The Wall Street Journal* in the autumn of 1994, involved three individuals (who had been writing letters and articles about the proposed housing project) and the Department of Housing and Urban Development (HUD) which had engaged in investigations of the three individuals and had sought fines (Leo 1994; MacDonald 1994). The *U.S. News & World Report* writer argued that 'Citizens should be free to engage in the routine push and shove of NIMBY politics, favoring one site over another, or protesting a neighborhood's oversaturation with facilities for addicts and the homeless' (Leo 1994: 20). NIMBY strategies by residents must be allowed, this writer continued to argue, because '[p]rojects to aid the down and out are producing too much social chaos' (Leo 1994: 20). Efforts by HUD to enforce the Americans with Disabilities Act and the Fair Housing Amendments Act were characterized by American Civil Liberties Union (ACLU) lawyers as impeding the rights of individuals to dissent from governmental action. Thus, in protecting residents' rights to engage in NIMBY tactics, the ACLU was working to redefine NIMBY strategies from exclusionary action to constitutionally protected dissent.

As this example indicates, fairness in the context of the NIMBY syndrome has most often centred on host communities. For example, the NIMBY syndrome is widely understood and addressed by policy-makers and service providers as the exclusion of controversial facilities from particular neighbourhoods. As discussed in previous chapters, to combat such exclusion state and federal legislation has increasingly been passed and implemented to prevent discrimination by residents, communities, and local governments against specific groups, although, as shown in the Berkeley example, with varying degrees of success.[1] However, beyond the desires of community members, fairness and justice in the context of the NIMBY syndrome must also concern the needs and desires of facility users, who navigate the existing and often fragmented network of service providers and government agencies to address a myriad of needs.

To explore the dynamic nature of fairness inherent in the NIMBY syndrome, this chapter outlines the dimensions of what I term community-centred and user-centred perspectives. I will illustrate how these varying perspectives of fairness and justice are contextual, with the perception of injustice as defined through a community-centred perspective often serving to activate locational conflict across the US. Overall, I argue the community-centred perceptions of unfairness in facility siting has widened across class, race, and location, creating a new and still fragmented conceptualization of what constitutes fairness in siting human service facilities. Similarly, there have been shifts in user-centred perceptions in fairness, sparking new and distinct (re)presentations of homelessness and HIV/AIDS in US society. This chapter describes the elements of this problematic of fairness within the NIMBY syndrome, contrasting these two perspectives.

With a community-centred perspective, various substantive and procedural issues concerning fairness and justice have constituted the mainstay of scholarly and policy debate. Such issues have included the criteria by which fairness can be defined and evaluated (using neighbourhood or municipal boundaries often coupled with socio-demographic characteristics such as race or class), the procedures and decision-making processes by which such strategies are developed and implemented which have left many communities feeling marginalized and powerless, and the strategies by which such criteria have been used to construct fair siting outcomes.

When considering the problematic of fairness in the NIMBY syndrome and human service delivery, however, any definition and characterization must also centre on the just and fair treatment of facility users. From this perspective, the definition of fairness in service delivery has most often centred on issues of access. Such issues have included the geographic location and operation of facilities, the advantages and disadvantages of service clustering, and the desires and perceptions of service-dependent individuals in determining appropriate locations and types of services.

To illustrate the conceptual points made, I will draw primarily from one high visibility instance where municipal governmental agencies, community leaders, and residents were central actors in determining what was fair and just in terms of the location of homeless persons and facilities: the case of New York City's Fair Share siting criteria applied to twenty-four proposed homeless shelters. At issue in this case were considerations of the appropriate and inappropriate location of homeless persons and facilities, the acceptable level of input by communities and municipalities in excluding homeless persons and facilities, and the extent of community and municipal responsibility for homeless persons located within their socially and legally defined boundaries.

When community-centred and user-centred perspectives of fairness collide in practice, planners, researchers, and political decision-makers are often left with the difficult question of determining the differential legitimacy of these conflicting claims (whether vocal or not) to develop appropriate siting strategies. The chapter thus concludes by exploring some of the issues in balancing the needs of communities and facility users.

Community-Centred Perspectives

Debates concerning fairness usually focus on the lack of fairness given specific substantive outcomes and processes (following Smith 1994).[2] This section will focus on the dimensions of inequitable siting outcomes and marginalizing decision-making procedures to explore the reasons why individuals and municipalities perceive a lack of fairness or justice in terms of human service facility siting.

In terms of outcomes, researchers, advocates, and community activists have increasingly argued that specific neighbourhoods have hosted disproportionate numbers of controversial facilities as the result of inequitable siting outcomes. Historical inequities in siting outcomes have created a spatial map where specific communities have relatively large concentrations of undesirable and controversial facilities, while others enjoy the use of more desirable facilities and amenities. However, facility siting procedures have also emerged as a focal point for debates concerning fairness. Specifically, researchers and community residents have argued that decision-making processes may marginalize individuals, making them feel powerless over land-use policies and siting outcomes. Such powerlessness may spark local opposition as a way to countermand the unresponsiveness by decision-makers perceived by community members.

Inequitable Siting Outcomes

As David Smith (1979) has argued, 'The geographical perspective [on equity] may be summarized as a concern with who gets what *where*, and how' (p. 18, emphasis in text). But with respect to the NIMBY syndrome, the question of inequity in siting outcomes relates more closely to the question of who does not get what, where, and how. That is, when dealing with controversial, and usually undesirable, human service facilities, assumed not to be demanded by local residents (an issue I will return to later in the chapter), the important concern is why specific communities bear a greater burden of hosting controversial facilities when compared to others.

Some of the most nuanced and contextualized discussions about inequity and controversial land use have emerged from the environmental justice and environmental hazards siting literature, particularly in geography. Rather than thinking about inequality and justice solely in terms of the statistical and spatial relationships between race and the location of controversial sites, such research has shown that these observable relationships are indicative of embedded social and spatial structures. According to Laura Pulido (1996*a*), for example, the central relationships in issues of environmental justice as social movement consist of 'structured and institutionalized inequality' (p. 4). If we use this as a starting-point for discussing inequitable siting outcomes and human services, one central dimension of such inequality in terms of human service delivery is the stigmatization of persons and places associated with homelessness and HIV/AIDS.

As I discussed in Part II, the stigmatization of persons and places has conditioned the spatial concentration of controversial human service facilities (especially identified with homelessness and HIV/AIDS), and resulted in the organized opposition of diverse communities directed against the proposed siting and operation of such facilities in close proximity to residences and businesses. The expression of stigmatization has meant that the spatial distribution of facilities has been geographically uneven, with concentrations tending to emerge and becoming maintained in Skid Rows and inner-city areas of metropolitan areas (Wolch 1996).[3] As Michael Dear and Jennifer Wolch (1987) argue in their book on homelessness, 'as the assignment unfolded [of clients to facilities, and of facilities to neighbourhoods], host communities that became service-dependent ghettos were heavily burdened relative to other communities. Thus one of the most severe infringements of client and community rights arose as a result of highly uneven and eventually inequitable distributions of clients among communities in metropolitan regions' (p. 226).

The concern about facility overconcentration and saturation is highly dependent on the localized perception of residents about the number of facilities that defines 'too many'. In specific neighbourhoods, 'too many' may mean one facility (whether it is a homeless shelter, a group home for persons living with HIV/AIDS, or some other type of human service facility), while for other communities, there is a tipping-point at which the alarms about facility saturation are sounded throughout the neighbourhood. In the upscale community of Dana Point (in Orange County, California), for example, three board-and-care homes for elderly persons did not appear to alarm residents, but with the proposal for two additional group homes (housing between three and five elderly persons), neighbourhood residents began a petition drive to prevent the opening and operations of these additional homes (Takahashi and Dear 1997). Arguments revolved about the inappropriateness of the neighbourhood for elderly persons, and how the community should not be forced to house an undue burden in terms of human services in the region.

Facility saturation for many communities is less an issue of defining how many is 'too many', and more an issue of a large and greater number of facilities when compared to other communities. There are neighbourhoods, such as Skid Rows and many inner-city communities in large metropolitan areas, which have historically shouldered a greater share of the burden for siting human service facilities (providing welfare, homelessness, mental health, and substance abuse services) (Wolch 1996). As the need for facilities has risen over the past decade, the disparity in spatial concentrations of human service facilities has become increasingly unacceptable to those neighbourhoods carrying more than 'their "fair share" of the caring burden' (Dear and Wolch 1987: 25).

The siting of human services in Skid Rows and inner-city areas has been conditioned in part by the cost of facility siting, and by organized opposition in largely privileged neighbourhoods. Inner-city areas and Skid Rows have often represented the least expensive option for service providers and local governments for new construction or rehabilitation of buildings for use as human service facilities. The low cost of land and buildings, and concomitant low market value, stems in large degree from local and extra-local market forces and metropolitan restructuring, which have depressed land and housing values in specific neighbourhoods. Such processes of devaluation both reflect and condition institutional and structural inequality (following Pulido 1996a) and define social and spatial stigmatization with respect to facility siting.

A related dimension to the issues of facility saturation and low cost land in specific communities is the perception that the siting of a facility attracts additional facility users and facilities, a so-called 'magnet effect' (Rowe 1997). There are two dimensions to such a magnet effect: the potential attraction of additional facility users (i.e. homeless persons and people living with HIV/AIDS) to neighbourhoods and municipalities playing host to a human service facility; and the attraction of additional facilities to a neighbourhood or municipality once a facility or a number of facilities is sited locally (i.e. the fear of facility saturation).

The first dimension of a magnet effect consists of facilities being widely perceived to attract individuals who desire access to services to areas proximate to such facilities. The problem associated with an increasing population of persons who might use or want to use the facility in question stems from the suspicion that more individuals will be drawn to the area, from geographical locations further and further from the neighbourhood, especially if the facility is effective in providing services. Clientele using facility services may be deemed to be less and less familiar as the distance they travel to the facility grows.

The second dimension of the magnet effect is the concern that additional facilities will be sited in a neighbourhood after one or more have been sited and are operating. The primary concern of residents and businesses in this dimension of the magnet effect is that once the door is opened for facilities to be sited, the inherent difficulty in siting controversial facilities elsewhere will mean that planners and service providers will increasingly turn to siting facilities in communities which have already accepted one (or several). Thus, the acceptance of one (or several) facilities is often associated with the potential (and probable) saturation of the neighbourhood with more and perhaps even different (but equally controversial) types of facilities.

Inner-city and older urban areas have often been the recipients of numerous and varying controversial facilities. The less equal character of inner-city and older urban areas (following Smith 1979) has also been reinforced through visible and organized neighbourhood opposition to human service facilities. Such actions have discouraged the siting of facilities outside inner-city and Skid Row areas of downtown metropolitan areas, since opposition has been largely associated with home-owning, suburban, and, though not often stated, White communities (Dear and Taylor 1982). Both to minimize the monetary cost of locating and constructing available sites for facilities, and to avoid the confrontation and conflict associated with siting controversial human services in these communities, planners and service providers in the past sought neighbourhoods of lesser resistance, which tended to consist of persons of colour, low-income residents, and renters (Smith and Hanham 1981; Dear and Taylor 1982).

When considering historically disadvantaged communities, discussions of inequity have more often centred on the needs of communities which are not addressed. This is perhaps the more familiar discussion in scholarship concerning the relationship between social justice and space, in struggles over the (re)distribution of resources and capital (e.g. Castells 1977; Harvey 1973). Human services associated with health care have long been lacking in low-income neighbourhoods and communities of colour (Kodras and Jones 1990; Smith 1979). But also of critical concern is the widespread need, particularly in low-income communities of colour, of general services and businesses, such as shopping malls and grocery stores (e.g. Takahashi 1997*b*). Historically disadvantaged communities may therefore perceive a double jeopardy in terms of controversial human services siting. Not only may communities have to play host to human services which they deem controversial, undesirable, and not addressing local needs, but they may also suffer a lack of facilities and services for which they have expressed a need.

Such issues are clear in a city planner's comments about community input during the development of New York City's fair share criteria:

The city's poorer neighbourhoods were particularly irate that they were being saturated by overconcentrations of facilities like shelters because most city-owned property [and therefore the least costly for the city to obtain and use] was in low-income areas. If cost alone were the siting determinant, these areas would continue to be preferred sites. On the other hand, we heard from community leaders that the city too often overlooked their needs for facilities to serve their neighborhoods or to support the economic development of regional commercial centres. (Weisberg 1993: 94)

Fairness in Procedure

What this quotation from a city planner also hints at is the importance of the process of facility siting decision-making in understanding fairness and the NIMBY syndrome. Much of the scholarly discussion about fairness, justice, and the NIMBY syndrome has centred on the uneven outcomes of inequitable siting practices.[4] Thus, because low-income neighbourhoods, communities of colour, and inner-city enclaves have historically borne an unequal share of facilities considered controversial or undesirable by the populace at large, such facility and land-use distributions over space have been characterized as unfair. But while the inequitable outcomes of facility siting constitute a vital dimension of fairness in the NIMBY syndrome, equally important are the procedural dimensions defining facility siting (following Smith 1994).

Scholarly, policy, and public debates on locational conflict over human service facility siting have for the most part centred on inequitable spatial distributions of facilities (e.g. Wolch and Dear 1993). Lessons for understanding fairness in the context of the NIMBY syndrome centring on procedural issues may be drawn from the expanding scholarship on environmental racism and environmental justice which has generally been concerned with procedural justice. In her widely cited book, Iris Young (1990) puts forth this argument in which explanations and solutions to inequality broaden from solely the (mal)distribution of wealth to assess 'the social structure and institutional context that often help determine distributive patterns' (quoted in Pulido 1994: 916). For scholarly explorations of environmental racism, this characterization of environmental justice as procedural in nature has, as Laura Pulido (1994) has argued, meant that struggles have been framed as 'making the process of environmental decision-making more open and accessible to all people, especially marginalized communities' (p. 915).

In his work on locational conflict over hazardous waste, Robert Lake further argues that the scholarly and policy emphasis on community opposition towards facility siting is an outcome of how the issue of hazardous waste management itself is defined, that is, that the definition of hazardous waste management as a siting issue creates the locational conflicts defined as the NIMBY syndrome. In other words, rather than looking to outcomes of facility siting (such as concerns over risk or equity), Lake and his co-authors have instead defined locational conflict over hazardous waste siting 'in terms of the structural constraints that dispose the state to define management problems as siting problems and to arbitrate the siting disputes by means of interest-group conflict' (Lake and Disch 1992: 663). By focusing solely on outcomes, Lake and Disch (1992) further argue that overlooked is the 'fundamental structure of hazardous waste regulation' (p. 664). As they argue about conventional approaches to analysing justice in facility siting (p. 665):

Both an analytic approach and a policy response that remain at the level of outcomes obscure the prior question of why community residents are asked to accept a universally unwanted facility. More specifically, they neglect to ask why the hazardous waste management problem is inevitably presented to the public as a *locational* problem. Facility siting is actually the last step in a complex and multifaceted regulatory process, but it is universally the stage at which the clash of interests rises to the surface and where conflict is most readily apparent (emphasis in text).

From this previous scholarship, the rise of the NIMBY syndrome directed at human services can be conceptualized as stemming from a lack of fairness in procedure. Specifically, community opposition may be characterized as marginalization or oppression through procedure. That is, as communities are called upon to become hosts for human services, their notification that siting is being considered or proposed is often seen and characterized by these residents as their effective exclusion from the facility siting process, and, moreover, as having had a decision forced upon them without their participation or approval.

When residents understand facility siting processes as overlooking their views and concerns, they tend to feel marginalized by a process over which they have no control and which is designed to surround them with stigmatized individuals. Thus, even if planners and service providers may expect rejecting behaviours from communities with the proposed siting of human service facilities because of the societal stigmatization of homelessness and HIV/AIDS, people's actual negative reactions to stigmatized individuals and facilities may be further compounded and exacerbated by what residents perceive as an authoritarian process that marginalizes and oppresses them by denying their autonomy. When residents perceive themselves as marginalized and oppressed in the process of facility siting, the notion that homeless persons and people living with HIV/AIDS are stigmatized and marginalized by society may have little impact on their opposition. That is, the material marginalization of homeless persons or people living with HIV/AIDS becomes less important than the imagined and material marginalization that residents perceive that they themselves have experienced. This imagined and material marginalization occurs when residents are presented with a facility siting proposal over which they had little or no input, and, furthermore, when they are portrayed (by the popular media, researchers, or policy-makers) as unwilling to shoulder their fair share of societal problems. Education and awareness strategies to overcome the NIMBY syndrome (focusing on reducing negative attitudes towards homelessness and HIV/AIDS) may then have little effect if residents believe that they constitute the marginalized and oppressed population (rather than homeless persons or people living with HIV/AIDS).

The Promise of Fair Share

Fair share policies have been increasingly seen by policy-makers and service providers as a promising strategy for addressing not only historical inequity in facility siting and perceived marginalization by past decision-making processes, but also for minimizing future conflict. Simply put, fair share strategies entail the design and implementation of an even distribution of controversial facilities across neighbourhoods and larger political boundaries.

The growth in the need for human services over the past decade, particularly associated with homelessness and HIV/AIDS, has made the disparity in the spatial concentrations of human service facilities less acceptable for neighbourhoods where facilities are operating. Although zoning ordinances have been used in the past to try to correct the inequity of such spatial concentrations through minimum distance requirements between facilities, these have not been entirely effective in addressing the municipal and regional dimensions of facility saturation. As Michael Dear and Jennifer Wolch (1987) argue, 'It was only a matter of time before the cry to "open up the suburbs" was heard and the fair-share zoning or zoning "as-of-right" movement was born' (p. 25); they advocate a regional fair share approach to facility siting which would 'involve a change in the distribution of the spatial externalities associated with public facilities and specific populations in order to increase the equity of their incidence pattern' (p. 229).

Dear and Wolch (1987) believe that fair share policies have numerous advantages beyond the redressing of existing spatial inequities in facility siting. They believe that fair share policies are not solely remedial strategies, because they also address some of the fundamental sources of inequality 'by adjusting the base distribution of resources' (p. 230). In addition, they argue (along with Baer 1983) that fair share policies can be developed in adaptable, flexible, and politically appealing ways. Dear and Wolch lay out a schema for using the existing mechanism of zoning laws both to encourage the siting of human services across communities and at the same time to minimize the potential for facility saturation:

* clearly outline the characteristics of residential facilities within a specific size range appropriate for residential scales; for facilities which fall within this definition, provide 'by-right' siting procedures in single family residential zones

* similarly outline the characteristics for facilities larger than the specified size range, defining them as 'multifamily' and provide 'by-right' siting procedures within multifamily residential zones

* for non-residential facilities, such as drop-in centres, delimit such uses as commercial, permitting their location in areas already zoned for medical and legal offices

* define limits of spatial concentration for residential, multifamily, and commercially-defined human service facilities

* develop and implement health and safety regulations used in similar zoning classifications (adapted from Dear and Wolch 1987: 214)

But one of the significant challenges in designing and implementing fair share policies will always remain the determination of the fair number and content of facilities to be sited in specific communities. One recent example of the challenges to instituting fair share practices is New York City's recent effort to site homeless shelters across municipal borough boundaries.[5]

In 1989, the City revised its Charter and mandated the City's Planning Commission to construct criteria which would be 'designed to further the fair distribution among communities of the burdens and benefits associated with city facilities, consistent with community needs for services and efficient and cost-effective delivery of services and with due regard for the social and economic impacts of such facilities upon the areas surrounding the sites' (quoted in Weisberg 1993: 94). The Charter further required the City to develop a list of facilities which were scheduled or planned to be sited over the following two years, and an inventory of municipal and City-leased properties. The fair share criteria were adopted by the Commission in December 1990 and placed into effect in July 1991, after input from the City's fifty-nine community boards and five borough presidents. Barbara Weisberg (1993), an assistant executive director of the New York City Department of City Planning, argued that at its core the development of fair share criteria centred on planning, 'early and open consultation with communities,' and trying to consider both equity and efficiency concerns using lay (rather than expert) language (p. 94).

Is Fair Share Fair?

While the process for determining New York City's fair share criteria, and the criteria themselves, have been looked to by planners across the nation as a model for dealing with inequitable siting issues and the growing NIMBY phenomenon, critics have argued that the criteria did not fundamentally change the distribution of resources or facility burdens (e.g. Rose 1993). Although activists, especially those associating themselves with the environmental justice social movement, had hoped that the development of these criteria would provide an opportunity for addressing more structural inequalities, less stringent requirements were developed which the Planning Commission believed would be easier to implement. For example, New York City's fair share criteria did not forbid the siting of additional facilities in districts identified as having high concentrations of facilities, but did 'require closer scrutiny of the effects of facility clustering on neighborhood character' (Weisberg 1993: 96). In essence, 'the Planning Commission ultimately retreated to a less rigid requirement that city agencies merely consider the negative effects that concentrations of city facilities in particular neighborhoods might yield' (Rose 1993: 98).

And beyond criticisms about the lack of specific redistribution mechanisms within the criteria themselves, community activists also argued that the fair share criteria would do little to disperse controversial facilities geographically. The first time the criteria were used to site facilities, according to Joseph Rose (1993), executive director of the Citizens Housing and Planning Council in New York City, they 'nearly ignited an urban civil war' (p. 98). The conflict arose when Mayor David Dinkins, in October 1991, announced plans to build and operate 'twenty-four small, service-intensive residential treatment centers dispersed throughout the city' (Rose 1993: 98). The City of New York's Human Resources Administration (HRA) had outlined in two plans produced in January 1988 a strategy for eliminating large congregate family shelters and private sector hotels, such as armouries (described as Tier I), and for developing a set of new semi-private family shelters (Tier II) by 1992 (Gaber 1996). The HRA had estimated that the existing need for housing for homeless individuals and families required 3,800 new Tier II shelters, and planned to implement this strategy 'in co-operation with and without disruption of the concerned communities' (quoted in Gaber 1996: 308). The new construction required by this recommendation would enable an 'equitable geographic distribution of the facilities among the boroughs' (HRA 1988 quoted in Gaber 1996: 309).

Given the fair share criteria emphasis on placing facilities in neighbourhoods not already hosting multiple services, City officials proposed siting these shelters in thirty-five sites in neighbourhoods characterized as 'less saturated' (Rose 1993: 98). But the argument that this distribution was fair and based on criteria which were developed in collaboration with community members did little to minimize community outcry over the unacceptability of hosting homeless shelters. Joseph Rose (1993: 98) describes the public's response:

Residents collected petitions, staged irate protests, and leveled charges that the siting plan constituted a deliberate assault on the city's stable communities. Racial tensions increased and accusations and epithets about reactions to the plan dominated city politics. City officials and newspaper columnists derided the plan's opponents for their racist NIMBY sentiments, while neighborhood politicians claimed that the plan was concocted for political retribution. Since the siting procedures had been proclaimed with assertions of moral significance and buttressed by demands for equity, the opponents responded in kind. They charged that their communities were being unfairly singled out by calculations that did not consider the impact of other types of objectionable facilities and complained that more prosperous districts were being overlooked. Despite insinuations by the city that much of the opposition was racially motivated, minority and integrated neighborhoods targeted to receive shelters proved just as hostile to the plan as did the mostly white areas.

Therefore, although looked to as a solution for community-based opposition, oppressive decision-making procedures, and historical inequities in facility siting, the fair share criteria did little to change fundamental distributions of resources. As Sharon Gaber (1996) argues, '[a]lthough some shelters for single adults and families had opened in middle- and upper-income communities, most shelters still were located in lower-income communities' (p. 309). In New York City, the principles of fair share garnered wide public and political support; however, when put into use (i.e. when the criteria were used to site homeless shelters) neighbourhoods returned to the practices associated with NIMBY and protection of their turf (in this case defined through borough boundaries). Thus, while in the abstract the notion of fair share distributions of facilities associated with homelessness (and HIV/AIDS) appears to be the most popular and most promising strategy to correct past inequities in facility siting (and more generally of resource distribution), concrete implementation of such notions remains problematic, at times ineffective, and sometimes even highly provocative.

User-Centred Perspectives

What the previous discussion highlighted was the common focus on the welfare of communities and municipalities in discussions of fairness and the NIMBY syndrome. While fairness (and the lack thereof) in the siting of controversial human services is often constructed in public debate as the placement of facilities away from specific, often vocal, communities, one overlooked dimension in such dialogue is the characterization of fairness as it relates to the users of such facilities (i.e. homeless persons or people living with HIV/AIDS). The lack of inclusion of facility users in the calculus of the NIMBY syndrome has been both reflective of and contributing to their social and spatial stigmatization, and ultimately to their lack of standing and legitimacy in locational conflicts. In excluding user perspectives and needs, the NIMBY syndrome has been effectively used to manage the siting of facilities, to concentrate facilities in specific communities and jurisdictions and to prevent them from locating in others.

Questions concerning fairness in facility siting should consequently also centre on the stated needs and identified obstacles to services by facility users. Stated needs by persons who access or want to access services are not commonly used in facility siting disputes; however, such data are often collected for needs assessments and programme evaluations. In a recent survey of homeless persons in largely suburban Orange County, California, for example, respondents who believed their homelessness was due to financial disruption, such as unemployment or family break-up, expressed little need for counselling or medical services (Crane and Takahashi 1997). Instead, these respondents requested help in obtaining better jobs, job training, affordable housing, and money to pay the first month's rent. Stated preferences for help in accessing medical care among these respondents seemed to have little to do with financial considerations. Thus, in this sample, medical and financial needs appeared to be somewhat unrelated.

These and other survey data indicate that there are distinct subgroups within the populations who are homeless or living with HIV/AIDS, each of whom has different wants and needs; this indicates that single strategies to address homelessness (or HIV/AIDS), such as solely improving accessibility to affordable housing, or solely providing employment or medical services, will neither be adequate nor effective. Thus, a mixture of services tailored to specific subgroups affected by homelessness and HIV/AIDS in particular places would appear necessary.

The NIMBY syndrome (along with broader locational politics over homelessness and HIV/AIDS) has created relatively well-defined service-dependent ghettos with large concentrations of human service facilities, usually in the Skid Rows of large metropolitan areas (Wolch 1996). From the point of view of the users of human service facilities, this has a significant impact in terms of access for persons needing such services. Skid Rows are not only perceived by the wider public as potentially dangerous and troubling places, but such perceptions also abound within the population of persons homeless and/or living with HIV/AIDS. For persons seeking access to human service facilities, given this widely held belief about the dangerousness of places such as Skid Rows, difficult choices must be made about the trade-offs between the potential risk of assault or violence in search of appropriate human services in Skid Row, and the potential access they might gain to the existing network of service providers and peers (Rowe and Wolch 1990).

In defining areas of facility concentration across space, the NIMBY syndrome effectively creates and reinforces a spatial map of service-rich and service-poor communities, which, for individuals seeking services, manages their daily routines and use of space (Dear *et al.* 1994). Service-rich and service-poor can apply across diverse communities (where wealthy communities may have service-rich environments because of the proximate location of museums, hospitals, and schools), but in terms of homelessness and HIV/AIDS, service-rich places are often neighbourhoods with little or no political influence to prevent the siting of controversial human services, and which present low-cost options for local governments and service providers.

Service-rich communities in terms of homelessness and HIV/AIDS therefore are the result of a centralized service strategy, where large institutions or a clustering of many smaller agencies provide services for homeless persons or people living with HIV/AIDS (Wolch 1996). The list of obstacles to accessing services is long and complex but may be mitigated by a centralized service strategy. For homeless persons and people living with HIV/AIDS, obstacles are bureaucratic (e.g. intrusive procedures inhibiting use and access to services, chronic undersupply of specific services), logistic (e.g. lack of transportation, variations in eligibility and ability to pay), cultural (e.g. language differences, understanding of homelessness and HIV/AIDS as deviance or illness), and individual (e.g. multiple diagnoses and issues, social disaffiliation, and distrust of service providers and government agencies).

There are clear benefits in clustering services for people needing human services. Geographical clustering of services (which has been described as ghettoization) provides opportunities for interaction both among service providers and facility users, which may facilitate the meeting of the multiple needs of users and the co-ordination challenges faced by service providers. Facility users may, because of the physical proximity of varied types of services in such service clusters or hubs, encounter enhanced access to the human services network. Information may be more readily available, user social networks may be promoted, and any social distance between facility users and service providers may be more easily bridged. As Dear *et al.* (1994) argue, a proliferation of such hubs throughout a metropolitan area would not only provide the benefits of physical clustering of services for facility users, but would also overcome the negative aspects of ghettoization (such as social and spatial stigmatization and potential victimization of facility users).

There are, however, some disadvantages to a service cluster strategy which may reduce fairness from a user-centred perspective. Although a varied set of services may exist in the service-rich cluster, the mix of services in particular neighbourhoods may not accommodate all or even most of the needs of homeless persons or people living with HIV/AIDS. The need for specific services not contained in the centralized service network will therefore ultimately require individuals needing human services to make multiple trips between facility sites to acquire needed services (Dear *et al.* 1994).

Community- Versus User-Centred Perspectives

There is an important link between the perception that homelessness and HIV/AIDS are outside local community concerns, and the subsequent impact perceived to be caused by facilities and facility users on neighbourhoods or municipalities. The concept 'facility user as outsider' indicates that community-centred and user-centred perspectives on fairness are often in direct conflict. In other words, the characterization of homelessness and HIV/AIDS as outside community interests (both in terms of community members being homeless or living with HIV/AIDS, and in terms of whether community members should become involved in the politics of homelessness and/or HIV/AIDS) implies that facilities and their users bring with them threats to the everyday routines and practices of community members.

These threats centre on the moral and material dangers to communities along the dimensions of stigma outlined in Chapter 4. Facilities associated with homelessness and HIV/AIDS bring with them the classification of non-productive, dangerous, and personally culpable, which may permeate through to nearby homes and places. In addition, individuals identified as homeless or living with HIV/AIDS are embodiments of non-productivity, dangerousness, and personal culpability, indicating to other neighbourhood residents that their lives and their property are at potential risk (with this risk clarified by sensationalist media accounts, including those of homeless persons murdering their mental health case workers, or individuals living with AIDS purposefully engaging in risky sexual practices in an attempt to infect others deliberately). Moreover, there are perceived moral dangers in associating with homelessness and HIV/AIDS, or in allowing homelessness and HIV/AIDS to permeate community boundaries. As described in Chapter 7, such dangers centre on the potential changes which might occur to family and community norm structures, given the presence of homeless persons or people living with HIV/AIDS.

Such perceived physical and moral threats to community structure and stability are embedded in and define questions of communal responsibility for individuals and groups not deemed to be part of the community. The NIMBY syndrome clearly expresses the view that residents and businesses, although they may believe that homelessness and/or HIV/AIDS are tragic conditions which should be addressed, do not concomitantly accept that facilities providing services should be sited nearby. Thus, while in the abstract homelessness and HIV/AIDS remain issues which garner public support for federal and state spending, and hence require societal response, they are still often considered to be of concern to individuals outside the confines of local communities.

The collision between community-centred and user-centred perspectives is clearly illustrated by the dilemma encountered when historically disadvantaged groups oppose the siting of human services directed towards homelessness and HIV/AIDS. When the NIMBY syndrome was perceived as solely the bastion of home-owning, suburban, and White neighbourhood residents, scholars and policy-makers could easily dismiss such practices as clearly oppressive and effectively exclusionary. But with the increasing participation of communities deemed marginalized across social, political, and economic dimensions, such a characterization has become more difficult to ascribe. Instead, the inherent oppression and inequity of the NIMBY syndrome has been made problematic as scholars and community activists have described such practices as elements of new social movements, as indicative of wider struggles to redistribute resources, and as ways of (re)claiming political power in the politics of location.

Oppression as manifested through inequitable siting outcomes and marginalizing procedures has led many historically disadvantaged groups, such as communities of colour and low-income neighbourhoods, to become immersed in siting conflicts over human services, making necessary a rethinking of the perhaps simplistic characterization of the NIMBY syndrome as consisting solely of White, suburban, home-owners. Traditionally marginalized groups (such as communities of colour, and low-income neighbourhoods) have increasingly become participants in NIMBY strategies to prevent the siting of human services. At first glance, such participation may seem an empowering action which works to challenge oppressive political, economic, and social structures by (re)claiming control over local land use. In organizing to prevent the siting of what are deemed to be undesirable human services, however, traditionally marginalized communities may exclude other oppressed groups (such as homeless persons and people living with HIV/AIDS). The question then arises about the relative importance of marginalization in deciding where controversial facilities should be placed. In particular, is the marginalization of homeless persons and people living with HIV/AIDS (expressed through NIMBY practices to prevent facility siting) acceptable if the socio-spatial exclusion is carried out by another marginalized group (e.g. residents of a neighbourhood defined by race and class)?

Disentangling this complex issue of relative marginalization might be enabled by framing the question of fairness in terms of socio-spatial relationships of oppression and domination (following Young 1990). Depending on the definition of spatial fairness, facility users, oppressed communities, or privileged neighbourhoods might prove the beneficiaries in facility siting conflicts. If fairness is defined solely from a community-centred perspective, then the NIMBY syndrome as practised by historically disadvantaged communities might be characterized as empowerment for oppressed communities, with the socio-spatial exclusion of homeless persons or people living with HIV/AIDS becoming a probable outcome. However, if fairness in the NIMBY syndrome is defined from a user-centred perspective, then such actions would be characterized as exclusionary no matter what the characteristics of the host community (i.e. privileged or oppressed).

The possible negative outcomes of using solely a user-centred perspective on fairness might be that historically disadvantaged communities would remain excessively burdened by controversial facilities, as communities of privilege use legal and informal means to maintain social and spatial stigmatization. If historically disadvantaged groups (such as communities of colour and low-income neighbourhoods) cannot socio-spatially exclude other groups (such as homeless persons and people living with HIV/AIDS) through such means as the NIMBY syndrome, privileged groups may continue to find loopholes in laws and planning practices, thereby effectively protecting themselves from the siting of controversial facilities. But beyond the unacceptability of marginalizing facility users to benefit specific community residents, the exclusion of human services by historically disadvantaged communities may not resolve historical inequities in facility siting. That is, the likely result might be that the NIMBY syndrome would spread across all communities, and historically disadvantaged groups might become embroiled in conflicts with one another over facility siting (if current NIMBY practices are any indication), defining the conflict as one of neighbourhood versus neighbourhood rather than as a regional or extra-local issue.

What is clear from this discussion of the changing dynamics of the NIMBY syndrome is that opposition to persons and facilities, whether by dominant or disadvantaged neighbourhoods, is oppressive to persons homeless or living with HIV/AIDS. But what is also clear is that NIMBY practices by historically disadvantaged communities (e.g. low-income neighbourhoods and communities of colour) cannot be seen only from a user-centred perspective of fairness. Such practices must also be understood from a community-centred perspective of fairness and justice, challenging oppressive processes and structures which have created inequitable siting outcomes specifically and uneven development more generally.

Conclusion

There is little doubt that debates concerning the fairness (or lack thereof) in siting human services associated with homelessness and HIV/AIDS will not dissipate in the near future. The ideal expressed in New York City's fair share criteria of dispersing the burden associated with siting homeless shelters across neighbourhoods and boroughs will be looked to increasingly by planners and service providers to address concerns of facility saturation, historical inequities in facility distribution, and broadening cries of 'not in my back yard' across diverse communities. To the extent that more communities are using NIMBY strategies to control local land use, fair share or other facility distribution plans will need to be more coherent and enforceable than imagined by past researchers and policy-makers. In a pragmatic sense, state and local governments are being asked to address multiple and conflicting public demands. They are being told to resolve local manifestations of homelessness and HIV/AIDS, redress past siting inequities, and respond to public concerns about the potential dangers associated with siting human services (i.e. that neighbourhoods are not appropriate sites for human services, and that these facilities will damage or threaten local amenities and property values).

It is apparent that state and local governments will respond in varied ways to these significant and conflicting demands. Much of the response will depend on local experience and the current political position of local government officials. Large metropolitan areas which have had to deal with large and growing populations of homeless persons have varied widely in their response. As shown in Chapter 8, the City of Santa Ana chose to address its growing homeless population with anti-camping ordinances which government officials hoped would keep homeless persons from entering or remaining within municipal boundaries (and especially from entering the Civic Center area of downtown). But, as the discussion in this chapter indicated, New York City, a municipality largely associated with the late twentieth-century's crisis of homelessness, chose instead to pursue a process of fair share, to address the growing need for homeless shelters, to respond to the historical inequities in facility siting, and to address public concern with decision-making procedures. The newly emerging environment of human service facility siting will therefore necessarily be geographically shaped, reflecting not only the local historical experience with homelessness and HIV/AIDS, but also situational and specific politics concerning the nature and stigma of homelessness and HIV/AIDS. This new environment may, however, express the same person-place stigma (resulting in service-dependent ghettos in Skid Row and other less desirable areas) even with growing interest on the part of planners and policy-makers in fair share and other alternative facility distribution policies. To put the issue most clearly, whereas NIMBY strategies were used by home-owning, suburban residents and other privileged communities against the dispersion of human service facilities in the interests of retaining quality of life and property values, the development of fair share procedures as indicated in New York City may well serve the same interests and élites.

Consequently, planners, policy-makers, and service providers must be careful in attributing too much to the potential corrective benefits indicated by fair share procedures. While vital to the redistribution of human services across diverse communities, their increasing significance is as much the result of the growing sophistication of individuals and communities engaged in NIMBY strategies and of local and state governments trying to balance multiple public demands as it is a positive move towards greater equity in the distribution of societal burdens. In this sense, the inability of cities to develop and implement fair share procedures in siting human services reflects both the promise and potential of such procedures to correct past inequities but also the imagined and material threat of person-place stigmatization to communities not already acting as human service facility sites.

While not operating at the local level, efforts to effect fair share results have been central in federal legislation such as the Americans with Disabilities Act of 1992 (which makes illegal discrimination against disabled persons in housing and employment in intent or outcome) and the Fair Housing Amendments Act of 1988 (which makes illegal the making unavailable or denying a housing sale or rental to persons because of mental or physical disabilities, including mental disability, alcoholism, and HIV/AIDS). While these legal mandates at the federal level have been very necessary to stem present and future exclusionary planning practices, by creating significant changes in the language of local policies surrounding housing and economic development, they have been less effective in altering the historically inequitable patterns of human service delivery and the expanding geography of homelessness and HIV/AIDS, especially in large metropolitan areas. For both sides of fair share policy development and implementation, a crucial and often still missing element is the ability to enforce such redistributive policies in the face of vehement, vocal, and highly visible community opposition.

Underlying the arguments put forth about fairness in facility siting are fundamental questions about the nature of fairness and justice in US society. For example, I noted above that community-centred expressions of the NIMBY syndrome have to do with individual neighbourhoods and the element of fairness inherent in allowing them control (or not) over local land use rather than perceiving them as part of regions or states. This distinction is vital for several reasons. The clearest reason is that decisions about facility siting when NIMBY arises are applied on a neighbourhood-by-neighbourhood or municipality-by-municipality basis, and are not automatically reflective or useful to any particular conflict between a community and organizations wanting to site a facility (e.g. a planning department or a service provider). Less clear is that fair share policies at the local level treat the NIMBY syndrome as one that impacts on neighbourhoods and municipalities, rather than as one that impacts on homelessness and HIV/AIDS. This means that, given a fair share approach, the welfare of neighbourhood residents will ultimately be at the centre of disputes, implying that those living in communities should be privileged over persons affected by homelessness and HIV/AIDS.

In contrast, federal legislation (such as the Americans with Disabilities Act and the Fair Housing Amendments Act) making discrimination illegal in housing or employment broadens the discussion about the potential impact of human services, protecting neighbourhoods which have unduly been seen as sites of least resistance and spreading services across a municipality or region. In addition, such federal legislation identifies issues of fairness and justice with persons seen as 'disabled' (using a wide range of physical and mental disabilities as definitions), indicating a more user-centred perspective on what constitutes justice in society than that focused on communities as used in discussions of the NIMBY syndrome.

Fairness and justice cannot be seen as simply equal distributions across space when considering the problematic question of fair siting policies for controversial human service facilities. Because, along with the issues surrounding spatial location and the impact of facilities on residents and businesses, notions of fairness and justice must consider the circumstances of those homeless persons and people living with HIV/AIDS who might need and use human services. In specific cases fairness and justice might be served by an equal distribution of human services across space (e.g. a fair share strategy), by a geographic concentration of facilities and services for specific populations (e.g. a service hub policy), or the policy focus on eradicating the causes of homelessness and HIV/AIDS (e.g. the NIABY or Not In Anybody's Back Yard alternative; Heiman 1990). In any case, policy-makers and advocates must not only continue to design, implement, and, especially, enforce legislation, particularly at the federal level which makes illegal discrimination in housing, employment, and services access, but must also develop locally specific strategies and options to correspond to varied disputes over the definition and character of fairness in facility siting across diverse communities and client groups. These locally specific strategies will necessarily entail trade-offs among the various aims of spatial justice-oriented policies, and may include information dissemination (Takahashi 1998), negotiation and mediation (e.g. Susskind and Cruikshank 1987), compensation and incentives (e.g. Carnes *et al.* 1987; O'Hare and Sanderson 1993), and legislation and litigation (Dear 1992). The trade-offs will involve facility users, service providers, communities and neighbourhoods, municipalities, and regions, and must ensure that justice and fairness are not provided for one group at the expense of others, particularly those groups which have been socially and spatially stigmatized.

11

Facing the NIMBY Syndrome

THIS book began with the statement that growing need for human services has clashed with community response. To explain and illustrate this expanding conflict, I outlined the parameters of homelessness and HIV/AIDS in the US, and how the perspectives of community attitudes and social relations have been used effectively to define reasons for the longevity of the NIMBY syndrome. I added the dimension of social and spatial stigma to these perspectives to highlight further the deeply embedded nature of community opposition, using case studies to indicate how the NIMBY syndrome is both intractable and fragile in particular circumstances. Included in this discussion were analyses of geographical variation in community and municipal response to homelessness and HIV/AIDS, and the tensions involved in sorting out who is responsible (or, in the case of NIMBY, who is not responsible) for providing care and services to homeless persons and people living with HIV/AIDS. From statistical analyses of national attitudes through to case studies of municipal response to homelessness and the characterization of HIV/AIDS in communities of colour, I have worked to illustrate several diverse ways of understanding the connection between stigma and the NIMBY syndrome.

I want to reiterate that there is no single perspective or explanation which can fully articulate the reasons for the growth in the phenomenon I have described as the NIMBY syndrome. In this sense, I have not attempted to prove or disprove a singular hypothesis, nor have I tried to show how past conceptualizations are in error. Instead, what I have done in this book is to use various ways of conceptualizing and illustrating the connections between stigmatization and the NIMBY syndrome to provide distinct empirical perspectives on the dimensions of this expanding phenomenon. By so doing, I hope to have shed light on the problems faced by communities, the state, and human service providers in dealing with the contemporary issues of homelessness and HIV/AIDS. For example, a number of chapters indicated how the NIMBY syndrome might be perceived as an instance of empowerment (rather than exclusion), and thus should not be condemned outright as an expression of discrimination and oppression. In situations where planners and service providers must choose among communities in siting decisions, the definition of NIMBY as exclusion or empowerment and the relative success of NIMBY strategies (given existing social relations of resource distribution) might be paramount in eventual siting decisions. The tensions inherent in siting decisions make NIMBY an almost commonplace and even central aspect of current and future siting policies, as illustrated by the cases of homelessness in Los Angeles and Santa Ana.

In explaining the NIMBY syndrome as the social and spatial expression of stigma, I have steered away from any single conceptualization for analysing and addressing the NIMBY syndrome. A single approach to the NIMBY syndrome would indicate clearer explanations and policy strategies than the discussion I have presented. Explanations based in attitude structure, for example, argue that negative attitudes are the source of community opposition, and focus on information, education, and community input and participation as remedies for the NIMBY syndrome. While such perspectives indicate a clear set of strategies and rationale, I have argued in this book that they tend to overlook the complexity underlying the dynamism and diversification of this phenomenon, and may therefore indicate solutions that may be ineffective in specific circumstances.

Rather than pursuing an all-encompassing framework, I have used the case studies to show how the social and spatial constructions of stigma have varied across geographic locations, racial groups, and different conditions (i.e. homelessness and HIV/AIDS), and how these social and spatial constructions have been linked to localized response. In this chapter, I will continue this argument, that increasing locational conflict over human service facility siting cannot be solely attributed to self-serving, exclusionary neighbourhoods. I will suggest in this concluding chapter that the geography of the NIMBY syndrome should be understood as the intersection of three concomitant factors in time and space which have served to galvanize individual communities in opposition to human service facilities: structural and institutional factors which have economically, socially, and politically marginalized a growing proportion of the nation's population; the short-term intractability of the social crises of homelessness and HIV/AIDS; and the growing public dissatisfaction with planning departments and other federal, state, and municipal governmental institutions.

The Geography of the NIMBY Syndrome

A growing proportion of the nation's population has become marginalized. There is little doubt that homelessness and HIV/AIDS have reached staggering proportions particularly during the 1980s and early 1990s. The expansion and diversification of homelessness and HIV/AIDS have been widely commented upon since the 1980s. Early 1980s national estimates of the homeless population in the US, for example, ranged from a low of about 250,000 to a high estimate of approximately 3 million (Hombs and Snyder 1982; HUD 1984). But, by the 1990s, scholarly estimates had changed, emphasizing homeless episodes instead of homeless individuals, and increasing to between 840,000 for a given year (Wolch and Dear 1993) and 26 million over a period of five years (Link *et al.* 1994). HIV/AIDS saw similar patterns of increasing prevalence, particularly among communities of colour. These patterns and the changing definitions of homelessness and HIV/AIDS were discussed in Chapter 1.

In chapters throughout the book, three sources of rising public rejection and opposition towards homelessness and HIV/AIDS were analysed. First, homelessness and HIV/AIDS have become dispersed throughout metro-politan areas, and are now visible and therefore of public concern outside inner-city areas, in suburbs, and neighbourhoods which have not experi-enced homelessness or HIV/AIDS locally. The second source is related to the first. With the rise and dispersion of homelessness and HIV/AIDS, there has been a concomitant rise in the need for, and subsequent efforts to site, human service facilities. Community rejection (which may have been latent) has therefore become publicly expressed because of the growing need for, and subsequent planned siting of, human services associated with homelessness and HIV/AIDS. And the third source contributing to the contemporary rise in the NIMBY syndrome is the increased participation by municipal governments in finding legal ways to prohibit groups from entering and remaining in specific areas. Municipalities are becoming involved in legal actions both in the courts and in legislation to prevent homelessness from entering or remaining within jurisdictional boundaries.

When scholars, policy-makers, and pundits discuss the NIMBY syndrome, they often focus on the intractability of social issues such as homelessness and HIV/AIDS. The intractability of homelessness and HIV/AIDS implies to the wider public that human service facilities comprise only stopgap measures, that homeless persons and people living with HIV/AIDS will forever flow through these facilities, and that this flow is not liable to end in the near or foreseeable future. The futility, or at the very least the perceived lack of effectiveness, of public sector efforts to stem the tide of rising homelessness (and, to a lesser degree, the widening impact of HIV/AIDS) means that, no matter what communities do, homelessness and HIV/AIDS may continue unabated.

Throughout the book, I have emphasized that many communities and municipalities across race and class lines are becoming very hostile places for persons and facilities associated with homelessness and HIV/AIDS. Homeless persons and people living with HIV/AIDS are viewed as being unproductive, dangerous, and personally culpable for their marginalized positions in society. The destitution and illness often attributed to homeless persons or people living with HIV/AIDS, while often described as emanating from structural changes, are often still seen as directly emanating from deviant and abnormal behaviours or faulty decision-making on the part of those homeless or living with HIV/AIDS. As I have alluded to in previous chapters, the construction of homelessness and HIV/AIDS as unproductive, dangerous, and attributable to individual behaviour has provided the substance of conflict over the appropriate location of groups and facilities. Because homeless persons and people living with HIV/AIDS are believed not to be contributing members of society (i.e. they are drains on society), because they remain potentially dangerous (to life and property), and because even with mitigating circumstances they can still be seen as somehow to blame for their being and remaining homeless or HIV-positive, residents who see themselves as removed from these devalued dimensions work vigorously to keep homelessness and HIV/AIDS outside their neighbourhoods, thereby protecting themselves at least a little from the encroachment of expanding poverty, crime, and other social ills to which they want to believe they can be immune.

The case studies in this book which deal with these issues (in Chapters 7 and 8, for example) reflect the deep social rifts generated by these characterizations. While communities not currently experiencing homelessness and HIV/AIDS, and society at large, have much to gain by maintaining and promoting such stigmatized identities, the inability of planners, service providers, and homeless persons and people living with HIV/AIDS themselves systematically to counter the power of these negative labels and stereotypes has created a very unequal field in terms of adjudicating siting conflicts. The geography of the NIMBY syndrome is therefore more than the growing population of persons homeless or living with HIV/AIDS, and more than the self-centred, reactionary views of residents; it reflects perhaps a more fundamental dimension of contemporary society: the balkanization of society into the haves and the have-nots is forcing those perceiving themselves in the middle to side with the haves, clarifying their neighbourhood and municipal boundaries using visible demonstration, vocal opposition, legislation, and judicial decisions to demarcate where homelessness and HIV/AIDS are allowed to exist. This balkanization is contributing to the solidification not only of class identity (i.e. home-owning, gainfully employed, and suburban, or at least, not inner-city) but also of racial identity. The geography of community opposition is therefore reflective of the solidification and practice of multiple racial communities, including White but also Latino and African American.

Since the history of homelessness and HIV/AIDS is intricately connected to growth and change in particular communities, influenced by planner and policy-maker strategies and decisions for economic development and minimizing conflict with middle- and upper-class residents, and business owners and employees, this visible and vocal separation of community welfare from the welfare of marginalized groups has had devastating effects especially on solidarity within communities of colour. In one case study, I outlined the difficult choices faced by families and friends of homeless individuals and persons living with HIV/AIDS. Given such difficult circumstances it is difficult to judge families and friends for their attempts to maintain ties to the normal communities with which they identify themselves, even when they ostracize individuals homeless or living with HIV/AIDS. There are two results of such efforts at boundary maintenance. The maintenance of imagined community purity or clear community boundaries is one outcome. Another outcome, not just for communities of colour, but for all communities, is that, when faced with stringent definitions of who belongs to the community and who does not, individuals become less willing to join or follow household or community norms and rules regarding proper and acceptable behaviour and beliefs. Thus, the geography of the NIMBY syndrome should also be understood in terms of the efforts made by communities and municipalities to re-establish publicly the parameters of community and societal norm structures, including those characterized as racial and cultural.

But I cannot stress enough the fragility and fragmentation inherent in the stigmatization of homelessness and HIV/AIDS. While scholars and policy-makers, and I for the majority of this book, usually focus our attentions on the rejecting facets of societal response to homelessness and HIV/AIDS, the stigmatization of these two social crises has also spawned proactive efforts at inclusion and empowerment. In one case study documenting the federal, state, and local co-operative efforts at HIV education and prevention in the US, I outlined how a highly stigmatized condition such as HIV/AIDS, albeit when defined as a public health issue, may garner significant public sector funds and an emphasis on local information, implementation, and evaluation in implementing programmes and policies. After many years of national level awareness and prevention messages, the continued lack of education and understanding concerning HIV/AIDS especially within communities of colour (along with other institutional and political pressures) induced the Centers for Disease Control and Prevention (CDC), a federal agency, to mandate local-level planning and implementation. The stigmatization of HIV/AIDS was central to the new emphasis on local messages delivered by local messengers, and thus comprised a vital element to generating diverse and place-specific policies concerning HIV/AIDS. One of the elements of these local messages and messengers was the inclusion of persons living with HIV/AIDS and those populations identified to be at high risk. Thus, not only did this directive change the way in which policies were to be developed and implemented, it also changed the stakeholders and significant individuals who could design and evaluate policies and programmes. In this way, the stigmatization of HIV/AIDS served to bring those directly affected to the table of decision-making.

And, of course, during this era of rising NIMBY sentiments, scholars and policy-makers across the nation have worked to redefine homelessness and HIV/AIDS as national crises rather than individual or neighbourhood problems. The Fair Housing Amendments Act of 1988 and the Americans with Disabilities Act of 1992 were explicit statements that discrimination against persons mentally or physically disabled was not to be tolerated in any locale across the US. Scholars have further worked to redefine homelessness as reflective of broad structural changes in the economy, changing housing markets and demographic profiles, along with individual vulnerabilities placing many of us at risk of potential homeless episodes. In terms of HIV/AIDS, researchers and policy-makers have strived not only to find medical solutions for AIDS and for the transmission of HIV, but also to explore the prejudice towards HIV/AIDS and to discover ways to overcome individual, community, and societal discrimination. The Ryan White CARE Act funds are an expression of the understanding by policy-makers that addressing HIV/AIDS must not only centre on medical research, but also on changing individual and group behaviour concerning transmission. But while federal legislators and many researchers continue their struggle to redefine the character of homelessness and HIV/AIDS, prevailing attitudes concerning the undesirability and societal unacceptability of homelessness and HIV/AIDS continue to result in local opposition to human service facilities.

These prevailing attitudes have interacted with the eroding confidence of communities in the public sector's ability to address homelessness and HIV/AIDS, resulting in a widening NIMBY syndrome in the US. As the 1990s draw to a close, programmes and policies concerning homelessness and HIV/AIDS have not made significant strides in the public consciousness to eradicate homelessness and to address HIV/AIDS. Thus, efforts at redefining and dealing with homelessness and HIV/AIDS, while noteworthy and necessary, have not had a significant impact on the politics of human services siting. The public has grown weary of public sector attempts to use social engineering methods to fix what appear to be intractable social issues; and in the midst of these seemingly intractable problems, the public sector and politicians have become increasingly enmeshed in scandals and controversies which have further eroded their abilities (and the perception of their effectiveness) to deal with complex issues such as homelessness and HIV/AIDS. The eroding public confidence in public sector politicians and bureaucrats to address appropriately and effectively homelessness and HIV/AIDS has fuelled broad-based and ongoing efforts at welfare reform and the continuing eradication of any remaining safety net programmes at the federal, state, and local levels.

It is not surprising then that in this climate service providers, researchers, and policy-makers appear not to have made meaningful steps in addressing the widening popularity of the NIMBY syndrome, and, further, that residents practising NIMBY tactics have increasingly been characterized positively in the mass media as fighting an ever-growing and oppressive state, and promoting local autonomy.

Facing the NIMBY Syndrome

While I have argued throughout this book that NIMBY is the expression of the stigmatization of persons and places associated with homelessness and HIV/AIDS, I have also suggested in this chapter that the contemporary manifestation and rise of this phenomenon can be traced to the confluence of three interrelated factors: the structural and institutional changes which have marginalized a growing proportion of the nation's population; the perception that the social crises of homelessness and HIV/AIDS are intractable, at least in the near future; and the growing public dissatisfaction with public sector politicians and bureaucrats, over homelessness and HIV/AIDS, but more broadly as well. My argument that the geography of the NIMBY syndrome is the result of the confluence of these three factors at the end of the twentieth century rests on the core assumption that this current manifestation of public opposition differs from locational conflict documented since the turn of the century (e.g. Meyer 1995; Meyer and Brown 1989). The current rise of the NIMBY syndrome is the result of recent social, political, and economic events that have threatened communities and municipalities enough to express vocally and visibly their opposition to human services; this mobilization is in direct response to the three factors—the widening bipolarization of society across social, economic, and political dimensions; public recognition that homelessness and HIV/AIDS are long-term social issues likely to remain critical, and a concomitant compassion fatigue towards marginalized groups; and a widespread public disenfranchisement with institutional democracy. While we can hope that these conditions will improve, paving the way for lesser vocal and visible rejection of homelessness and HIV/AIDS, ongoing efforts at welfare reform and further attacks on redistributive policies, increasing bifurcation of wealth throughout the US, and growing protectionist sentiments in neighbourhoods and municipalities makes an optimistic prediction problematic.

One long-term method of minimizing the NIMBY syndrome is to reduce the reasons for why it has expanded. As Robert Lake (1993) has suggested, one fundamental requirement to completely eradicating NIMBY will be the elimination of homelessness and HIV/AIDS. While this may appear at first to be an academic solution to material social issues, scholars who study homelessness have indicated time and again that 'there is no great mystery about the steps necessary to eliminate the problem of homelessness in the United States' (Blau 1992: 180). Policy-makers, planners, advocates, and researchers agree that homelessness can be addressed through a combination of affordable housing with and without human services, employment with liveable wage levels and benefits, and accessible and appropriate social services providing needed help for a diverse population. HIV/AIDS is a more difficult situation because of the requisite medical care needed for the prevention of transmission and opportunistic infection, but many of the same issues of housing, services, and employment also apply. But while such proposed solutions appear obvious and even feasible, the concrete implementation of such notions is obstructed through changing political focus on the desirability of and need for a universal safety net, intergovernmental and interbureaucratic conflict concerning agency responsibility for current and proposed programmes, and, of course, localized opposition to implementation.

In the short term, while public and private efforts continue to pursue integrated and comprehensive strategies (such as HUD's Continuum of Care philosophy concerning homelessness), planners and service providers are left with the problem of how to address the NIMBY syndrome as the need continues and rises for facilities associated with homelessness and HIV/AIDS. To overcome the NIMBY syndrome (as it has become manifest in communities and municipalities), researchers and policy-makers have looked to legislation prohibiting discrimination and to the pre-emption of local land use control, education and awareness strategies attempting to allay the often (though not always) misinformed public concerning potential dangers and risks and also the potential benefits of having a human service facility nearby, and processes incorporating community input, participation, and negotiation (e.g. Dear 1992; Dear and Wolch 1987).

I have already discussed fragmented and conflicting objectives of intergovernmental legislative attempts to prevent discrimination (through the Americans with Disabilities Act and the Fair Housing Amendments Act), and to legalize the removal of homelessness from municipal boundaries (through anti-camping and anti-panhandling ordinances at the local level). The second set of strategies, those of education and awareness, offer an alternative and less coercive method of trying to achieve community and wider public acceptance of facilities. Information can be disseminated in multiple ways, and planners are devising various methods appropriate to communities which attempt to overcome the powerful influence of particular community members. Since research has indicated that the NIMBY syndrome often consists of small, vocal groups of residents, rather than a mass mobilization of neighbourhoods, the influence of these individuals tends to be overemphasized in facility siting processes. To downplay the influence of small, vocal groups, planners in Orange County, California, for example, are altering the ways of presenting information to community members prior to the initiation of environmental impact studies, site approval processes, and even the determining of zoning. Instead of more formalized processes where planners and/or developers of a site make formal presentations to a concerned public, these Orange County planners have decided that a less formal atmosphere would be more conducive not only to information dissemination but also to the management of individual community members who might tend to capture a meeting using a more formal input process (i.e. community members speaking one at a time at a microphone with questions and comments). A group of information tables with rotating community members provides a greater opportunity for one-on-one interaction between planners and residents, but also diffuses the possibility that one or several community members might take control of the meeting, setting the tone for later conflict, but also potentially inflaming the siting issue.

But when conflict does ensue over the siting of controversial facilities, negotiation and mediation have become popular methods to create consensus among a group of often very oppositional interests. Such methods focus on identifying and including important stakeholders, co-operative problem definition and development of alternatives, and the creation of plans which would respond to the interests expressed by stakeholders (Susskind and Cruikshank 1987). These methods have been very successful in bringing disparate and oppositional groups to consensual solutions concerning problems such as controversial facility siting. However, in such dispute resolution techniques, there is always the potential for subverting the empowerment of specific groups, through various methods ranging from the particular management of implementation (such as determining the times and locations for meetings enabling some and disabling others to attend, and by delineating the ability of the public to participate through agenda setting and other procedural rules and regulations) to broader control over discussion and access (such as through the use of rational-technical language and methods) (Lake 1994).

Further, the explanation of the NIMBY syndrome as socio-spatial stigmatization used throughout this book implies that while such negotiation and mediation techniques may be very successful in addressing facility siting conflicts as they occur, they only respond minimally to the underlying sources of such conflict. That is, although a siting conflict may be resolved in a particular neighbourhood, the social relations defining and reproducing socio-spatial stigma will mean that proposed human service facilities will likely face community rejection. Thus, as the group of those affected by homelessness and HIV/AIDS continues to grow, and as the need for and provision of human services continue to proliferate, the use of negotiation will become even more necessary.

Researchers and policy-makers have also proposed alternatives to formalized procedures for consensus building and the use of formal political institutions. Such researchers and policy-makers, such as Joel Blau (1992), have advocated grass-roots mobilization and coalition building (perhaps building upon the relative success of the environmental justice social movement) as a means of broadening political access and visibility, but also to create alliances with groups with greater access to resources (such as middle-income voters). Since mobilization by low-income and homeless persons alone is problematic at best for many reasons, including access to political influence and resources to sustain long-term activism (Piven and Cloward 1977), any successful and long-term mobilization will require a broad-based political coalition consisting of persons affected by homelessness and HIV/AIDS and middle-income populations.

If the NIMBY syndrome is any indication, such middle-income populations and communities can be mobilized when they perceive a direct threat to their livelihoods, quality of life, or property, or when they believe they are being stigmatized by an authoritarian and non-responsive decision-making process (specifically concerning facility siting, but also more broadly concerning homelessness and HIV/AIDS). However, building coalitions between low- and middle-income home-owners and homeless persons and/or people living with HIV/AIDS, rather than against them, may prove a formidable task. Such coalition building would require that low- and middle-income voters side with the have-nots in their vocal and visible concern over the distribution of facilities. There have been few instances where this has been the case. Instead, the stigmatization of homelessness and HIV/AIDS has resulted in our current context as the almost predictable occurrence of NIMBY. Although there are situations where the NIMBY syndrome is not the norm (for example, the Ventura River bed rescues and subsequent rapid co-ordination of services for homeless persons discussed in Chapter 6), contemporary social, economic, and political circumstances have created a context where boundaries (both physical and imagined) are being erected and fortified against the growing have-nots.

But while ongoing circumstances might indicate a pessimistic future in terms of addressing the NIMBY syndrome, homelessness, and HIV/AIDS, there continue to be individuals (homeless persons, people living with HIV/AIDS, and others), non-profit organizations, and planners and policy-makers who work tirelessly not only to face the NIMBY syndrome as it becomes vocal and visible, but also to unravel the seemingly intractable knot of social and spatial stigma. Many organizations involved in service provision are realizing that service provision without capacity building within affected populations and communities, and other efforts at changing the social and spatial identity of homelessness and HIV/AIDS, only provide short-term individual solutions to these two social crises. These efforts at capacity building, collaboration, and leadership development within the context of service provision have the potential for altering the professional bureaucratization of stigma. These and other ongoing efforts at realigning the identity and delineation of homelessness and HIV/AIDS may well serve as the fulcrum through which the stigmatization of persons and places may be more widely recognized, altered, and contested.

NOTES

Chapter 1

1. I use the term 'HIV/AIDS' to describe persons and places associated with HIV and/or AIDS. Researchers have argued, however, that HIV and AIDS are distinct biomedical and social categories. As Julia Epstein (1995) has suggested, 'AIDS requires as its *pre*condition an altered circumstance that is not itself a disease: HIV seropositivity. . . . With respect to social categories, persons with HIV infection are morally dangerous whereas persons with AIDS, while we do not cede their moral dangerousness, reside more thoroughly in the realm of the medicalized physically ill' (emphasis in text, p. 13).
2. There are various terms and labels used throughout the book which have encountered increasing problematizing by scholars and policy-makers, such as 'race', 'inner city' and 'gay'. For example, the term 'inner city' is laden with 'racial' and class stereotypes which make it a problematic term. The use of inverted commas around all these terms would be somewhat distracting, therefore I chose to discuss their social construction explicitly. I will discuss the socio-spatial stigmatization of such terms in Part II of the book.
3. The exclusionary and inclusionary facets of governmental response and the distinctive ways that homelessness and HIV/AIDS are addressed will be the focus of Chapters 8 and 9.
4. Barak (1991) provides a comprehensive description and classification of the changing definitions of homelessness.
5. Blau (1992) argues, however, that the actual proportion of homeless people who were formerly institutionalized belies the argument that deinstitutionalization is a significant cause of homelessness.
6. Rent control has also been cited as being a possible cause of housing shortages and consequently of homelessness (Tucker 1991). This argument and its evidence, however, have been disputed by Applebaum *et al.* (1991).
7. The literature on housing policy during the 1980s and early 1990s is far more extensive than can be included in this chapter. See Blau (1992) for a concise review of this literature as it pertains to homelessness.
8. The emphasis here is on strategies and programmes which address homelessness after individuals and families become homeless. There are also examples of programmes which work to prevent homelessness before households are actually forced on to the streets. Such programmes include the multi-state Homeowner-ship Protective Effort (HOPE), the Homelessness Prevention Program in New Jersey, the Rental Allowance Program in Maryland, and the Eviction Prevention/Rent Bank Program in Connecticut (Schwartz *et al.* 1991). Although these are few and seriously underfunded, they do provide examples where local and state governments have worked to prevent homelessness.

9. Fischer (1982) describes the form and function of social support networks.
10. 'Panhandling' describes activities by homeless person to elicit money from passers-by.
11. In particular, there has been a growing focus in research on homeless youth (e.g. Carlen 1996; Hutson and Liddiard 1994; Ruddick 1996). Youth homelessness exhibits both similarities and differences from homelessness among adults and families. While homeless youth face many similar risks and dangers to their mental and physical well-being as homeless adults, they are also seen as a distinctive group (termed 'runaways'), at times eliciting the public image of an adventurer (although this myth has been refuted by research).
12. While daily mobility is often limited, residential mobility may be very high due to the instability of shelter and housing within such populations (Dixon *et al.* 1993; Kearns and Smith 1993; Wolch *et al.* 1993).
13. The City of Santa Ana in Southern California, for example, recently passed and began implementing an anti-camping ordinance, in effect making sleeping outdoors an illegal and punishable activity.
14. There has been dispute among biomedical researchers, however, about whether HIV actually causes AIDS or not (e.g. Duesberg 1991).
15. The characterization of HIV/AIDS as epidemiology has encountered growing critique. Researchers have increasingly argued the centrality of the social construction of HIV/AIDS in understanding public response, scientific research, and public health policy (Brown 1995; Kearns 1996; Patton 1990, 1994).
16. There has been a growing emphasis in geography and other social sciences concerning the problematic nature of 'race' and the importance of the social construction of 'HIV/AIDS' in framing public response and public policy initiatives (e.g. Anderson 1991; Patton 1990; Pulido 1996*a*, *b*; Takahashi 1997*a*).
17. Because of this relative separation of homelessness and HIV/AIDS, the remaining chapters also treat these two populations, conditions, and bases of stigmatization somewhat distinctly.
18. There is, of course, a large body of biomedical research exploring the nature and change in retroviruses with the eventual goal of developing a vaccine for HIV/AIDS prevention.
19. Chiotti and Joseph (1995), however, document the acceptance of a Toronto neighbourhood in the establishment of an AIDS hospice. There have been other documented examples of community support for the siting of AIDS-related human services or shelters (e.g. Vanderpriem 1996); however, these remain the exception rather than the norm.

Chapter 2

1. Zoning has long been used as a way of managing immigrant populations. As Frank Popper (1981) points out, 'In the 1880s, San Francisco used zoning to prevent the spread of Chinese laundries' (p. 55).
2. In addition, Veness (1994) argues that homeless shelters may be more accept-

able to donors, public officials, and local residents if their overall goal is to mainstream individuals into permanent housing situations.

3. Ordered logit estimation is used here to accommodate the categorical nature of the dependent variables (i.e. acceptability of each facility measured on the six-point Likert scale). This method constructs a linear function of the independent variables and a set of *cut* points, which are then used to estimate an underlying score. The probability of respondents being accepting of each facility is determined by the probability that the estimated linear function is within the range of a specific group of cut points identified as corresponding to accepting attitudes. Rather than directly calculating the probability of acceptance or rejection, the sign of the statistically significant coefficients are analysed to determine the role of the location variables after controlling for other socio-economic and demographic characteristics.

Chapter 3

1. Fuss (1989) argues similarly about the differences in identity politics between gay men and lesbians, 'the stronger lesbian endorsement of [essentialist] identity and identity politics may well indicate that lesbians inhabit a more precarious and less secure subject position than gay men' (p. 98).

2. I will return to issues of justice and fairness concerning community opposition in Chapter 10.

3. I will discuss in more depth the intersection of race, HIV/AIDS, and community response in Chapters 5 and 7.

4. Although I am not attempting to argue in this book that community opposition towards environmental hazards can also be explained and explored using a framework based in the social relations of stigma, there have been recent scholarly attempts to make this claim, providing a conceptual bridge between the NIMBY syndrome as it relates to both human services and environmental hazards (see, for example, Gregory *et al.* 1996). Lake (1993) has also constructed a conceptual framework for understanding locational conflict over both environmental hazards and human services using a production relations approach.

5. Peter Marcuse (1988) has made the argument that the visible rise in homelessness in the US has garnered so much media and public attention specifically because it indicates that social and moral limits are disintegrating (cited in Wagner 1993: 10–11).

6. The sociological study of deviance is not limited to the exploration of stigma. The sociology of deviance, following the work of Émile Durkheim and Talcott Parsons, encompasses a broad set of theories and substantive focus, including psychological and psychoanalytic perspectives, physiology and biology, anomie concepts, theories of cultural transmission, functional strain theories, and class conflict (e.g. Ellis 1987; Sagarin 1975; Traub and Little 1985).

7. Social psychologists have argued that stereotypes, labelling, and stigmatization can be understood through three dimensions: socio-cultural, psychoanalytic, and cognitive. According to Jones *et al.* (1984), the socio-cultural dimension

indicates that stigma is developed from the information we receive about homelessness and HIV/AIDS and our direct personal experiences, the psychoanalytic
or motivational dimension indicates that stigmatization is an ego-defensive
strategy which 'represents the projection of thoughts or impulses that the
attributor wishes to deny' (p. 161), and the cognitive dimension indicates that
stigma results from our cognitive processes (the ways that particular categories
and labels come to mind, how we associate 'unusual or distinctive stimuli' with
homelessness and HIV/AIDS, and how easily we can imagine ourselves as
homeless or living with HIV/AIDS). The explanation of stigma presented here
builds on the socio-cultural dimension outlined by social psychologists, and
emphasizes the ways that stigma is produced and reproduced through social
and spatial relations.
8. See Wagner (1993) for a discussion of the changing nature and significance of the
work ethic in US society.

Chapter 4

1. As Mary Douglas (1966) in her seminal work on pollution argues concerning the
constitution of primitive religions, 'As we know it, dirt is essentially disorder.
There is no such thing as absolutely dirt: it exists in the eye of the beholder. . . .
Dirt offends against order. Eliminating it is not a negative movement, but a
positive effort to organise the environment' (p. 2). Michel de Certeau (1986) has
also commented on the connection between the Other and rottenness, in particular, how rottenness is not only associated with the Other, but also with
institutions and practices controlling the Other and discourses surrounding
the Other.
2. Frances Fox Piven and Richard Cloward (1982) have argued that, more generally, restructuring of the welfare state and reduction of state regulation of
business reflect attempts by the state and capital interests to increase business
profits.
3. Risk groups for homelessness include households living at or near the poverty
level, those experiencing domestic violence, engaging in substance abuse, and
working in minimum-wage, contractual, and unstable employment. At risk
groups for HIV/AIDS transmission include persons engaged in unprotected
sex (usually male to male sexual contact), intravenous/injecting drug users,
and haemophiliacs.
4. There is also a large literature on the social production of space, in large part
building on concepts developed by Henri Lefebvre.
5. My goal here is not to present an in-depth description of the rich history of Skid
Row, Los Angeles, but rather to indicate the broad dimensions that have contributed to its creation as a place-myth. For a recent and expansive discussion of
the economic, social, environmental, and political trends which have influenced
the social and material construction of Los Angeles, see Scott and Soja (1996).
6. The notion of 'imagined geographies' has been discussed in much greater depth
by Gregory (1994), Lefebvre (1991), and Soja (1989).

7. Mary Douglas (1966) has argued that 'To have been in the margins is to have been in contact with danger, to have been at a source of power' (p. 97).

Chapter 5

1. There is of course a substantial and growing body of scholarship which has worked to problematize the notion of race and to explore the racialization of persons and places, especially in relation to undesirable land uses and facilities (e.g. Almaguer 1994; Hurley 1995; Jackson and Penrose 1993; Oboler 1995; Omi and Winant 1994; Pulido 1996*a*, *b*; Small 1994).
2. For a historical view of the participation of women in US social life, see Hewitt and Lebsock (1993).
3. The overrepresentation of African Americans (and the underrepresentation of Hispanics) among studied homeless populations is tempered however by gender. According to Burt's (1992) national study, Latina and African American women with children tend to be overrepresented among those homeless.
4. Injecting drug use is also a major route of transmission for HIV/AIDS within the homeless population (Nyamathi 1992).
5. Vulnerability to sexual assault and physical battery are exacerbated by mental disability in homeless women (e.g. Goodman *et al*. 1995).
6. This builds on the notion that persons of colour are carriers of race (in contrast to 'White' identity as devoid of race) and women are carriers of gender.
7. This argument is very similar to those used by geographers exploring the social and spatial relations of gender and sexuality with respect to female sex workers, gay/lesbian populations (e.g. Forest 1995; Hubbard forthcoming; Valentine 1993).
8. Biomedical research has also explored racial differences in HIV drug therapies (Cotton 1991). Indeed, much research on HIV/AIDS has focused on the importance of race and ethnicity as predictors or indicators of HIV infection and risk (e.g. Hu *et al*. 1995).
9. The linkages between male sex workers/prostitutes and HIV/AIDS transmission is not well documented, but published reports indicate the very high rates of HIV/AIDS within this population, especially among adolescent males (e.g. Markos *et al*. 1994; McKeganey 1994).

Chapter 7

1. 'Latino' and 'Latina' are used in this chapter to describe men and women of Latin American descent. In Orange County, California, these individuals are mostly of Mexican descent, born either in the US or in Mexico. The persons interviewed self-identified themselves as members of the 'Latino' community.

2. For a more general discussion of the perceptions of other racial groups, such as African Americans, towards HIV/AIDS and homelessness, see Chapter 5.
3. The proportion of African Americans in Orange County increased slightly over the past decade, but remained relatively small at approximately 39,000 in 1990 (approximately 2 per cent).
4. This epidemiological portrait of Orange County is drawn largely from the 'Plan for the Prevention of HIV in Orange County, California: 1996–1999' (Orange County HIV Prevention Planning Committee 1995).
5. There was a decrease in the number of deaths between 1992 (N = 310) and 1993 (N = 243).
6. I will return to the HIV Prevention Planning Process in Orange County, California, in Chapter 9.
7. The beliefs of popular individuals, for example, have been shown in a study of gay men in three small southern cities in the US to influence significantly the beliefs and behaviours of other gay men concerning high risk behaviour (Kelly *et al.* 1992). In addition, a recent national survey of attitudes towards controversial human services suggested that community leader endorsement is a significant predictor of accepting attitudes towards mental health care facilities (Takahashi 1998).
8. As one of the Vietnamese respondents stated, 'The fact that anyone who could speak English well enough to answer all these questions already throws you off the curb' (Vietnamese male, 40s, executive director housing/human services). This constraint emphasizes the care that must be taken in extrapolating the results to the entire Vietnamese or Latino communities in Orange County.
9. This respondent asked about the nature of such facilities and was told that they include buildings such as hospices, apartments, or group homes which people living with HIV/AIDS might use.

Chapter 8

1. As discussed in Chapter 1, the connections among public policy and the NIMBY syndrome will be treated separately for homelessness (Chapter 8) and HIV/AIDS (Chapter 9). Because the public, service providers, and policy-makers treat homelessness and HIV/AIDS in very distinct ways, Chapters 8 and 9 illustrate the varying ways that municipal governments have responded in exclusionary and inclusionary ways.
2. The discussion in this section is based primarily on Stoner (1995).
3. A widely cited and examined case is *Callahan* v. *Carey* in New York, where Robert Hayes, a lawyer in private practice, filed a class-action lawsuit arguing that individuals in New York had a right to shelter. In early December 1979, 'the New York Supreme Court granted the homeless men a preliminary injunction requiring the City to provide sufficient beds for the homeless men who applied for shelter' (Hopper and Cox 1986: 309; also Blau 1992). This decision was drawn from existing provisions of the New York Constitution, the State Social

Services Law, and the City's Administrative Code, and resulted in the temporary order that the City had to provide shelter to anyone who applied for shelter at the Men's Shelter in New York City (Hopper and Cox 1986).

4. *Rubin et al.* v. *City of Santa Monica*, CV 93–1255 LGB, U.S. Dist. Ct., 1993. The City of Santa Monica had developed an ordinance 'regulating the use of public parks by requiring groups of 35 or more to obtain a permit before using any park' (Stoner 1995: 156). The plaintiffs, who were homeless activists, were represented by American Civil Liberties Union lawyers in the US District Court of California suit.

5. Challenging municipal and state laws on the grounds that they burden the constitutional right to travel has long been used (see Ades 1989 for a review).

6. Studies of homelessness in Orange County indicate that the homeless population tends to be more dominated by Caucasians/Anglos when compared to inner-city metropolitan homeless populations, although, similar to studies of homelessness across the nation, racial minorities tend to be overrepresented in the Orange County homeless population. According to a survey conducted in 1993 of homeless persons seeking services at public and private, non-profit agencies, the homeless population was primarily Caucasian/Anglo (56 per cent), with 24 per cent of the respondents Latino/Hispanic, and about 13 per cent African American (Crane and Takahashi 1997). When compared to the 1990 US Census, Caucasians/Anglos were underrepresented among Orange County's homeless population (compared to 65 per cent of the County's population) and African Americans were overrepresented (compared to 2 per cent of the overall population). This survey also indicated that most homeless individuals were male (62 per cent), unmarried (56 per cent were single), and had lived in the County for long periods of time (58 per cent had lived in the county for ten years or longer). Homeless persons seeking services were primarily among the ranks of the new homeless, with approximately 72 per cent of those surveyed reporting being homeless for less than one year.

Chapter 9

1. The Ryan White Comprehensive AIDS Resources Emergency (CARE) Act of 1990 supported programmes providing out-patient and ambulatory medical and other services for people living with HIV/AIDS; 'Services in those urban areas with the highest number of reported AIDS cases were authorized under Title I of the act, formula grants to all States under Title II, and early intervention services under Title III. Title IV authorizes general activities, including pediatric research initiatives' (Bowen *et al.* 1992: 492).

2. SAMHSA provides funding to local health jurisdictions for HIV counselling, testing services, and other therapeutic treatments to persons who are in substance abuse treatment programmes.

3. The Education and Prevention Services branch is also responsible for overseeing the HIV Testing Program, Training Program, and HIV Early Intervention Program. The HIV Testing Program provides funds to local health

departments and community clinics for HIV testing and counselling services. The $10 million allocated for this programme in 1994 was distributed through formalized agreements or contracts with 130 entities to provide counselling and testing services at 400 sites statewide. The Training Program provides direct training services to State Office of AIDS contractors and other state and local agencies. The HIV Early Intervention Program funds twelve local Early Intervention Programs statewide (one of which is in Orange County), with the goals of interrupting HIV transmission to uninfected persons and improving the lives of persons living with HIV/AIDS (California Community Planning Working Group 1995).

4. Housing Opportunities for Persons With AIDS (HOPWA) is a Department of Housing and Urban Development (HUD) programme, providing funding for housing and services for low-income persons living with HIV/AIDS and their families. Beginning in 1993, the largest city in an EMSA (Eligible Metropolitan Statistical Area) was designated as the grantee for the HOPWA formula funds on behalf of the metropolitan area.

5. Primary prevention strategies are used to prevent infection from occurring in uninfected persons. Secondary prevention provides programmes to prevent or delay the onset of symptoms, illness, or disease in infected persons.

Chapter 10

1. There have been important city- and state-level policy efforts in redistributing the burdens or negative impacts widely seen to be caused by controversial human service facilities. States have continued to use their pre-emptive powers over land use to site facilities considered controversial by communities, although with limited success. At the federal level, legislation such as the Americans with Disabilities Act and the Fair Housing Amendments Act continue to be used to override local land-use zoning and conditional use permit processes which have in the past been extremely successful in preventing the siting of group homes and other types of human service facilities.

2. Susan Fainstein (1997) provides a critical review of three perspectives on social justice and urban space, focusing on political economic views, poststructuralist perspectives, and urban populist formulations of social justice, and Low and Gleeson (1997) provide a concise summary of the growing literature on conceptualizing justice particularly in the context of the postmodernist critique of universalist frameworks (also Merrifield and Swyngedouw 1997).

3. The notion of uneven distribution of services, and of persons needing those services, has long been discussed in geography and planning (Kirby 1982; Wolch and Gabriel 1985).

4. Researchers have also argued that characterizing fairness solely by the (un)even distribution of facilities and hazards does not capture the nuanced and contextual dependence of the concept as it is practised. Keller and Sarin (1988, also 1995) argue, for example, that 'an unequal distribution of risks may be preferred if it is accompanied by a compensatory differential in benefits

consistent with peoples' preference tradeoffs between received benefits and assumed risks' (p. 135).

5. A detailed description of New York City's bureaucratic procedures concerning public facility siting and the use of the fair share criteria is included in Valetta (1993).

BIBLIOGRAPHY

ADES, P. (1989), 'The Unconstitutionality of "Antihomeless" Laws: Ordinances Prohibiting Sleeping in Outdoor Public Areas as a Violation of the Right to Travel', *California Law Review*, 77: 595–628.

AGNEW, J. A. (1987), *Place and Politics: The Geographical Mediation of State and Society* (Boston: Allen & Unwin).

AINLAY, S. C., and CROSBY, F. (1986), 'Stigma, Justice, and the Dilemma of Difference', in S. C. Ainlay, G. Becker, and L. M. Coleman (eds.), *The Dilemma of Difference: A Multidisciplinary View of Stigma* (New York: Plenum Press), 17–38.

AIRHIHENBUWA, C. H., DiCLEMENTE, R. J., WINGOOD, G. M., and LOWE, A. (1992), 'HIV/AIDS Education and Prevention Among African-Americans: A Focus on Culture', *AIDS Education and Prevention*, 4: 267–76.

ALMAGUER, T. (1994), *Racial Fault Lines* (Berkeley: University of California Press).

ALRECK, P. A., and SETTLE, R. B. (1985), *The Survey Research Handbook* (Homewood, Ill.: Irwin).

ALTMAN, D. (1994), *Power and Community: Organizational and Cultural Responses to AIDS* (New York: Taylor & Francis).

ALVAREZ, F. (1995), 'Floodwaters Sweep Away Inertia', *Los Angeles Times*, 30 January, A3, A16.

—— (1996) 'Up from the Bottom', *Los Angeles Times*, 16 January, A3, A17.

AMARO, H. (1995), 'Love, Sex, and Power: Considering Women's Realities in HIV Prevention', *American Psychologist*, 50: 437–47.

—— and HARDY-FANTA, C. (1995), 'Impact of Ryan White CARE Act Title I on Capacity Building in Latino Community-Based Organizations: Findings from a Study of Two Cities', Report prepared for the US Department of Health & Human Services, Public Health Service (Washington, DC: Health Resources and Services Administration, Bureau of Health Resources Development).

ANASTOS, K., and MARTE, C. (1991), 'Women—The Missing Persons in the AIDS Epidemic', in N. F. McKenzie (ed.), *The AIDS Reader: Social, Political, and Ethical Issues* (New York: Meridian/Penguin), 190–9.

ANDERSON, B. (1983), *Imagined Communities: Reflections on the Origin and Spread of Nationalism* (London: Verso).

ANDERSON, K. (1991), *Vancouver's Chinatown: Racial Discourse in Canada, 1875–1980* (Montreal: McGill-Queen's University Press).

—— and GALE, F. (1992) (eds.), *Inventing Places: Studies in Cultural Geography* (Melbourne, Vict.: Longman Cheshire).

ANDERSON, N. (1923), *The Hobo* (Chicago, Ill.: University of Chicago Press).

Anonymous (1987), 'Group Homes No Threat', *Planning*, 53: 34.

—— (1990), 'Homeless Go-Around' (Editorial), *Los Angeles Times*, 29 August, B6.

—— (1991), 'Santa Ana Will Pay Homeless to Settle Lawsuits in Arrests', *New York Times*, 22 August, A18.

Anonymous (1995), 'New Orleans Considers Anti-Camping Law to Deter Homeless', *New York Times*, 26 November, I21.

—— (1996) 'More Pressures on the Homeless', *America*, 174: 3.

APPLEBAUM, R. P., DOLNY, M., DREIER, P., and GILDERBLOOM, J. I. (1991), 'Scapegoating Rent Control: Masking the Causes of Homelessness', *Journal of the American Planning Association*, 57: 153–64.

ARANDA-NARANJO, B. (1993), 'The Effect of HIV on the Family', *AIDS Patient Care*, 27–9.

ARNO, P. (1991), 'Housing, Homelessness, and the Impact of HIV Disease', in N. F. McKenzie (ed.), *The AIDS Reader: Social, Political, and Ethical Issues* (New York: Meridian/Penguin), 177–89.

ARRANDALE, T. (1993), 'When the Poor Cry NIMBY', *Governing*, 36–41.

ASCH, A. (1986), 'Will Populism Empower Disabled People?' in H. C. Boyte and F. Reissman (eds.), *The New Populism: The Politics of Empowerment* (Philadelphia: Temple University Press), 213–30.

AUSTERBERRY, H., and WATSON, S. (1986), *Housing and Homelessness* (London: Routledge and Kegan Paul).

BAER, W. (1983), 'Towards a Theory of Fair Share', mimeo, School of Urban and Regional Planning (Los Angeles: University of Southern California).

—— (1986a), 'Housing in an Internationalizing Region: Housing Stock Dynamics in Southern California and the Dilemmas of Fairshare', *Environment and Planning D: Society and Space*, 4: 337–50.

—— (1986b), 'The Shadow Market in Housing', *Scientific American*, 255: 29–35.

BAHR, H. (1973), *Skid Row: An Introduction to Disaffiliation* (New York: Oxford University Press).

BAILEY, M. E. (1991a), 'Developing a National HIV/AIDS Prevention Program Through State Health Departments', *Public Health Reports*, 106: 695–701.

—— (1991b) 'Community Based Organizations and CDC as Partners in HIV Education and Prevention', *Public Health Reports*, 106: 702–9.

BAKER, S. G. (1994), 'Gender, Ethnicity, and Homelessness: Accounting for Demographic Diversity on the Streets', *American Behavioral Scientist*, 37: 476–504.

BALDASSARE, M. (1990), 'Suburban Support for Non-Growth Policies: Implications for the Growth Revolt', *Journal of Urban Affairs*, 12: 197–206.

—— and KATZ, C. (1994), 'Orange County Annual Survey: Final Report' (University of California-Irvine).

—— —— (1995), 'Public Attitudes Towards AIDS Prevention and Education', sponsor report prepared for the Irvine Health Foundation and FHP Foundation, 1995 Orange County Annual Survey, University of California, Irvine.

—— and WILSON, G. (1995), 'More Trouble in Paradise: Urbanization and Decline in Suburban Quality-of-Life Ratings', *Urban Affairs Review*, 30: 690–708.

BARAK, G. (1991), *Gimme Shelter: A Social History of Homelessness in Contemporary America* (New York: Praeger).

BASSUK, E. L. (1987), 'Feminization of Homelessness: Families in Boston Shelters', *Community Mental Health Journal*, 26: 425–34.

—— (1991), 'Homeless Families', *Scientific American*, 265: 66–75.

—— (1993), 'Social and Economic Hardships of Homeless Women and Other Poor Women', *American Journal of Orthopsychiatry*, 63: 340–7.

—— and RUBIN, L. (1987), 'Homeless Children: A Neglected Population.' *American Journal of Orthopsychiatry*, 52: 1–9.

—— and ROSENBERG, L. (1988), 'Why Does Family Homelessness Occur? A Case-Control Study', *American Journal of Public Health*, 78: 783–8.

BAUM, A. S., and BURNES, D. W. (1993), *A Nation in Denial: The Truth About Homelessness* (Boulder, Colo.: Westview Press).

BAUMOHL, J. (1989), 'Alcohol, Homelessness and Public Policy', *Contemporary Drug Problems*, 16: 281–300.

BEAUREGARD, R. A. (1995), 'If Only the City Could Speak: The Politics of Representation', in H. Liggett and D. C. Perry (eds.), *Spatial Practices: Critical Explorations in Social/Spatial Theory* (Thousand Oaks, Calif. Sage Publications), 59–80.

BECKER, H. (1963), *Outsiders: Studies in the Sociology of Deviance* (New York: Free Press).

BELCHER, J. R., and DIBLASIO, F. A. (1990), 'The Needs of Depressed Homeless Persons: Designing Appropriate Services', *Community Mental Health Journal*, 26 3: 255–66.

BENNETT, A., and SHARPE, A. (1996), 'AIDS Fight is Skewed by Federal Campaign Exaggerating Risks', *The Wall Street Journal*, 1 May, A1.

BERG, L. D., and KEARNS, R. A. (1996), 'Naming as Norming: "Race," Gender, and the Identity Politics of Naming Places in Aotearoa/New Zealand', *Environment and Planning D: Society and Space*, 14: 99–122.

BERGER, P. L., and LUCKMANN, T. (1966), *The Social Construction of Reality: A Treatise in the Sociology of Knowledge* (New York: Doubleday).

BLASI, G. (1994), 'And We Are Not Seen: Ideological and Political Barriers to Understanding Homelessness', *American Behavioral Scientist*, 37: 563–86.

BLAU, J. (1992), *The Visible Poor: Homelessness in the United States* (New York: Oxford University Press).

BLENDON, R. J., and DONELAN, K. (1989), 'AIDS, the Public, and the "NIMBY" Syndrome', in D. E. Rogers and E. Ginzberg (eds.), *Public and Professional Attitudes Toward AIDS Patients: A National Dilemma* (Boulder, Colo.: Westview Press), 19–30.

BLISSLAND, J. H., and MANGER, R. (1983), 'Qualitative Study of Attitudes Toward Mental Illness: Implications for Public Education', paper presented at the Annual Convention of the American Psychological Association.

BLUNT, A., and ROSE, G. (1994) (eds.), *Writing Women and Space: Colonial and Postcolonial Geographies* (New York: The Guilford Press).

BOGUE, D. J. (1963), *Skid Row in American Cities* (Chicago, Ill.: University of Chicago Press).

BONDI, L. (1993), 'Locating Identity Politics', in M. Keith and S. Pile (eds.), *Place and the Politics of Identity* (London: Routledge), 84–101.

BOWEN, G. S., MARCONI, K., KOHN, S., BAILEY, D. M., GOOSBY, E. P., SHORTER, S., and NIEMCRYK, S. (1992), 'First Year of AIDS Services Delivery Under Title I of the Ryan White CARE Act', *Public Health Reports*, 107: 491–9.

BOWSER, B. P., FULLILOVE, M. T., and FULLILOVE, R. E. (1990), 'African-American Youth and AIDS High Risk Behavior: The Social Context and Barriers to Prevention', *Youth & Society*, 22: 54–66.

BOYTE, H. C., BOOTH, H., and MAX, S. (1986), *Citizen Action and the New American Populism* (Philadelphia: Temple University Press).

BRACHO DE CARPIO, A., CARPIO-CEDRARO, F. F., and ANDERSON, L. (1990), 'Hispanic Families Learning and Teaching About AIDS: A Participatory Approach at the Community Level', *Hispanic Journal of Behavioral Sciences*, 12: 165–76.

BRATT, R., HARTMAN, C., and MEYERSON, A. (1986) (eds.), *Critical Perspectives on Housing* (Philadelphia: Temple University Press).

BROWN, M. (1995), 'Ironies of Distance: An Ongoing Critique of the Geographies of AIDS', *Environment and Planning D: Society and Space*, 13: 159–83.

BULLARD, R. (1990), *Dumping in Dixie: Race, Class and Environmental Quality* (Boulder, Colo.: Westview Press).

—— (1993) (ed.), *Confronting Environmental Racism* (Boston: South End Press).

BURT, M. B. (1992), *Over the Edge: The Growth of Homelessness in the 1980s* (New York: Russell Sage Foundation).

—— and COHEN, B. E. (1989a), *America's Homeless: Numbers, Characteristics, and Programs that Serve Them* (Washington, DC: The Urban Institute Press).

—— —— (1989b), 'Who is Helping the Homeless? Local, State, and Federal Responses', *The Journal of Federalism*, 4: 111–28.

BUSH, A. J., and BOLLER, G. W. (1991), 'Rethinking the Role of Television Advertising During Health Crises: A Rhetorical Analysis of the Federal AIDS Campaigns', *Journal of Advertising*, 20: 28–37.

CAIN, R. (1995), 'Community-Based AIDS Organizations and the State: Dilemmas of Dependence', *AIDS and Public Policy Journal*, 10: 83–93.

California Community Planning Working Group (1995), 'California HIV Prevention Plan' (San Francisco: Harder+Kibbe Research).

CALSYN, R. J., and MORSE, G. (1990), 'Homeless Men and Women: Commonalities and a Service Gender Gap', *American Journal of Community Psychology*, 18: 597–608.

CARLEN, P. (1996), *Jigsaw—A Political Criminology of Youth Homelessness* (Buckingham: Open University Press).

CARNES, S. A., COPENHAVER, E. D., SORENSEN, J. H., SODERSTROM, E. J., REED, J. H., BJORNSTAD, D. J., and PEELLE, E. (1987), 'Incentives and Nuclear Waste Siting: Prospects and Constraints', in R. W. Lake (eds.), *Resolving Locational Conflict* (New Brunswick, NJ: Center for Urban Policy Research, Rutgers University), 353–75.

CARRIER, J., and MAGAÑA, J. R. (1992), 'Use of Ethnosexual Data on Men of Mexican Origin for HIV/AIDS Prevention Programs', in G. Herdt and S. Lindenbaum (eds.), *The Time of AIDS: Social Analysis, Theory, and Method* (Newbury Park, Calif.: Sage Publications), 243–58.

CASTELLS, M. (1977), *The Urban Question* (Cambridge, Mass.: MIT Press).

—— (1983), *The City and the Grassroots: A Cross-Cultural Theory of Urban Social Movements* (Berkeley: University of California Press).

CATON, C. L. M., SHROUT, P. E., EAGLE, P. F., OPLER, L. A. (1994), 'Risk Factors for Homelessness Among Schizophrenic Men: A Case Control Study', *American Journal of Public Health*, 84 2: 265–70.

Centers for Disease Control and Prevention (CDC) (1988), 'America Responds to AIDS Fact Sheet' (Atlanta, G.: Centers for Disease Control and Prevention), September.

—— (1996) 'Cumulative AIDS cases' (CDC World Wide Web home page) (Atlanta, G.: Centers for Disease Control and Prevention).

CHAN, S. (1989) (ed.), *Social and Gender Boundaries in the United States* (Lewiston, NY, The Edwin Mellen Press).

CHIOTTI, Q. P., and JOSEPH, A. E. (1995), 'Casey House: Interpreting the Location of a Toronto AIDS Hospice', *Social Science and Medicine*, 41: 131–40.

CHOW, R. (1993), 'Gone but not Forgotten', *Orange County Register*, 8 June, B1.

CLARK, G. L., and DEAR, M. (1984), *State Apparatus: Structures and Language of Legitimacy* (Boston: Allen & Unwin).

COCHRAN, S. D., MAYS, V. M., and LEUNG, L. (1991), 'Sexual Practices of Heterosexual Asian-American Young Adults: Implications for Risk of HIV Infection', *Archives of Sexual Behavior*, 20: 381–91.

COCKBURN, C. (1977), *The Local State: Management of Cities and People* (London: Pluto Press).

COHEN, B. E., and BURT, M. R. (1990), 'The Homeless: Chemical Dependency and Mental Health Problems', *National Association of Social Workers*, 3: 8–17.

—— —— and CHAPMAN, N. (1992), 'Food Sources and Intake of Homeless Persons', *Journal of Nutrition Education*, 24: 45–51.

COHEN, C. I., and SOKOLOVSKY, J. (1989), *Old Men of the Bowery: Strategies for Survival Among the Homeless* (New York: Guilford Press).

COLLINS, P. H. (1990), *Black Feminist Thought: Knowledge, Consciousness, and the Politics of Empowerment* (Boston: Unwin Hyman).

COTTON, P. (1991), 'Race Joins Host of Unanswered Questions on Early HIV Therapy', *Journal of the American Medical Association*, 265: 1065–6.

COUSINEAU, M. R., and WARD, T. W. (1992), 'An Evaluation of the S-Night Street Enumeration of the Homeless in Los Angeles', *Evaluation Review*, 16 4: 389–99.

COX, K. R. (1989), 'The Politics of Turf and the Question of Class', in J. Wolch and M. Dear (eds.), *The Power of Geography* (Boston: Unwin Hyman), 61–90.

CRANE, R., and TAKAHASHI, L. M. (1997), 'Who Are the Suburban Homeless and What do They Want? An Empirical Study of Stated Preferences for Assistance', mimeo, Department of Urban and Regional Planning (University of California, Irvine).

CRIMP, D. (1992) (ed.), *AIDS: Cultural Analysis/Cultural Criticism* (Cambridge, Mass.: MIT Press).

CROW, D. (1994), 'My Friends in Low Places: Building Identity for Place and Community', *Environment and Planning D: Society and Space*, 12: 403–19.

CUFF, D. (1989), 'The Social Production of Built Form', *Environment and Planning D: Society and Space*, 7: 433–47.

CUNNINGHAM, J. K., ROBINSON, G. L., and SERPE, R. T. (1993), 'Homeless Persons in Orange County: Demographics, Needs and Health-Risk Behaviors', Report prepared for the County of Orange, Health Care Agency, Document No. RDR-101 (Santa Ana, Calif.: County of Orange).

CURTIS, J. R., and PATRICK, D. L. (1993), 'Race and Survival Time with AIDS: A Synthesis of the Literature', *American Journal of Public Health*, 83: 1425–8.

DALTON, H. L. (1989), 'AIDS in Blackface', *Daedalus*, 118: 205–27.

DAUNT, T. (1994*a*), 'Outcry Prompts Mayor to Revise Homeless Plan', *Los Angeles Times*, 27 October, B1.

DAUNT, T. (1994*b*) 'Council OKs Plan for Homeless Center', *Los Angeles Times*, 9 November, B1.

—— (1994*c*) 'LAPD Begins Crackdown on Skid Row Homeless', *Los Angeles Times*, 19 November, B1.

DAVIDOFF, P. (1965), 'Advocacy and Pluralism in Planning', *Journal of the American Institute of Planners*, 31: 331–8.

DAVIS, D. (1991), '"Understanding AIDS"—The National AIDS Mailer', *Public Health Reports*, 106: 656–62.

DAVIS, M. (1985), 'Urban Renaissance and the Spirit of Postmodernism', *New Left Review*, 151: 106–13.

—— (1990), *City of Quartz* (New York: Verso).

—— (1991), 'Afterword—A Logic Like Hell's: Being Homeless in Los Angeles', *UCLA Law Review*, 39: 325–32.

DEAR, M. (1976), 'Spatial Externalities and Locational Conflict', in D. B. Massey and P. W. Batey (eds.), *Alternative Frameworks for Analysis* (London: Pion), 28–34.

—— (1988), 'The Postmodern Challenge: Reconstructing Human Geography', *Transactions of the Institute of British Geographers*, 1–13.

—— (1992), 'Understanding and Overcoming the NIMBY Syndrome', *Journal of the American Planning Association*, 58: 288–301.

—— GABER, S. L., TAKAHASHI, L., and WILTON, R. (1997), 'Seeing People Differently: The Socio-Spatial Construction of Disability', *Environment and Planning D: Society and Space*, 15: 455–80.

—— and GLEESON, B. (1991), 'Community Attitudes Toward the Homeless', *Urban Geography*, 12: 155–76.

—— and LAWS, GLENDA (1986), 'Anatomy of a Decision: Recent Land Use Zoning Appeals and Their Effect on Group Homes Locations in Ontario', *Canadian Journal of Community Mental Health*, 51: 5–17.

—— and TAYLOR, S. M. (1982), *Not On Our Street: Community Attitudes Toward the Mentally Ill* (London: Pion).

—— and WOLCH, J. (1987), *Landscapes of Despair: From Deinstitutionalization to Homelessness* (Princeton: Princeton University Press).

—— —— (1994), 'Herding the Homeless is an Unjust Answer' (Commentary), *Los Angeles Times*, 14 November, B7.

—— —— and WILTON, R. (1994), 'The Service Hub Concept in Human Services Planning', *Progress in Planning*, 42: 173–271.

DE CERTEAU, M. (1986), *Heterologies: Discourse on the Other*, trans. Brian Massumi (Minneapolis: University of Minnesota Press).

DE LAURETIS, T. (1990), 'Eccentric Subjects: Feminist Theory and Historical Consciousness', *Feminist Studies*, 16: 115–50.

Department of Health Services (1995), 'Guidelines for HIV Prevention Community Planning for Local Health Departments' (Sacramento, Calif.: California State Office of AIDS).

—— (1996*a*), 'HIV Prevention Community Planning Year One Review Team Report Orange County' (Sacramento, Calif.: State Office of AIDS).

—— (1996*b*), 'AIDS Case Registry: HIV/AIDS Reporting System Surveillance Report' (Sacramento, Calif.: State Office of AIDS).

D'ERCOLE, A., and STREUNING, E. (1990), 'Victimization Among Homeless Women: Implications for Service Delivery', *Journal of Community Psychology*, 18: 141–51.

DES JARLAIS, D. C., and FRIEDMAN, S. R. (1994), 'AIDS and the Use of Injected Drugs', *Scientific American*, 270: 82–8.

DEUSTCHE, R. (1990), 'Men in Space', *Artforum*, 28: 21–3.

DIAZ, T., BUEHLER, J. W., CASTRO, K. G., and WARD, J. W. (1993), 'AIDS Trends among Hispanics in the United States', *American Journal of Public Health*, 83: 504–9.

DIBLASIO, F. A., and BELCHER, J. R. (1992), 'Keeping Homeless Families Together: Examining Their Needs', *Children and Youth Services Review*, 14: 427–38.

—— —— (1995), 'Gender Differences Among Homeless Persons: Special Services for Women', *American Journal of Orthopsychiatry*, 65: 131–7.

DI RADO, A. (1994), 'Appeals Court Voids Santa Ana Ban on Camping by Homeless', *Los Angeles Times*, 4 February, A27.

DIXON, L., FRIEDMAN, N., and LEHMAN, A. (1993), 'Housing Patterns of Homeless Mentally Ill Persons Receiving Assertive Treatment Services', *Hospital and Community Psychiatry*, 44: 286–8.

DOLAN, M. (1994), 'Narrow Ruling is Expected on Homeless Law', *Los Angeles Times*, 9 December, A1, A37.

—— (1995), 'State Justices Uphold Tough Homeless Law', *Los Angeles Times*, 25 April, A1, A10.

DOUGLAS, M. (1966), *Purity and Danger: An Analysis of Concepts of Pollution and Taboo* (New York: Frederick A. Praeger).

DOWELL, D. A., and FARMER, G. (1992), 'Community Response to Homelessness: Social Change and Constraint in Local Intervention', *Journal of Community Psychology*, 20 1: 72–83.

DRAKE, R. E., OSHER, F. C., and WALLACH, M. A. (1991), 'Homelessness and Dual Diagnosis', *American Psychologist*, 46 11: 1149–58.

DREYFUSS, J. (1984), 'Silver Lake: Neighborhood of Tolerance', *Los Angeles Times*, 27 September, V1, V12–14.

DRUCKER, E. (1990), 'Epidemics in the War Zone: AIDS and Community Survival in New York City', *International Journal of Health Services*, 20 4: 601–15.

—— (1991), 'Drug Users with AIDS in the City of New York: A Study of Dependent Children, Housing, and Drug Addiction Treatment', in N. F. McKenzie (ed.), *The AIDS Reader: Social, Political, and Ethical Issues* (New York: Meridian/Penguin), 144–76.

DUESBERG, P. H. (1991*a*), 'Human Immunodeficiency Virus and Acquired Immunodeficiency Syndrome: Correlation but Not Causation', in N. F. McKenzie (ed.), *The AIDS Reader: Social, Political, and Ethical Issues* (New York: Meridian/Penguin), 42–73.

—— (1991*b*), 'Virus Hunting and the Scientific Method', *Science*, 251 4995: 724.

DURKHEIM, É., and MAUSS, M. (1963), *Primitive Classification*, trans. R. Needham (London: Cohen & West).

EARICKSON, R. J. (1990), 'International Behavioral Responses to a Health Hazard: AIDS', *Social Science and Medicine*, 31: 951–62.

EDGERTON, R. B. (1967), *The Cloak of Competence: Stigma in the Lives of the Mentally Retarded* (Berkeley: University of California Press).

ELKIN, S. L. (1987), *City and Regime in the American Republic* (Chicago: University of Chicago Press).

ELLIS, D. (1987), *The Wrong Stuff: An Introduction to the Sociological Study of Deviance* (Don Mills, Ont.: Collier Macmillan).

ENGLAND, K. V. L. (1991), 'Gender Relations and the Spatial Structure of the City', *Geoforum*, 22: 135–47.

EPSTEIN, J. (1995), *Altered Conditions: Disease, Medicine, and Storytelling* (New York: Routledge).

ERICKSON, J. and WILHELM, C. (1986), 'Introduction', in J. Erickson and C. Wilhelm (eds.), *Housing the Homeless* (New Brunswick, NJ: Center for Urban Policy Research, Rutgers University), pp. xix-xxxv.

FAINSTEIN, S. S. (1997), 'Justice, Politics, and the Creation of Urban Space', in A. Merrifield and E. Swyngedouw (eds.), *The Urbanization of Injustice* (Washington Square, NY: New York University Press).

FAUGIER, J., and CRANFIELD, S. (1995), 'Reaching Male Clients of Female Prostitutes: The Challenge for HIV Prevention', *AIDS Care*, 7: S21–S32.

FEAGIN, J. R. (1975), *Subordinating the Poor: Welfare and American Beliefs* (Englewood Cliffs, NJ: Prentice-Hall).

FERNANDO, M. D. (1993), *AIDS and Intravenous Drug Use: The Influence of Morality, Politics, Social Science, and Race in the Making of a Tragedy* (Westport, Conn.: Praeger).

FINK, P. J., and TASMAN, A. (1992) (eds.), *Stigma and Mental Illness* (Washington, DC: American Psychiatric Press).

FIRST, R. J., ROTH, D., and AREWA, B. D. (1988), 'Homelessness: Understanding the Dimensions of the Problem for Minorities', *Social Work* (Mar/Apr): 120–4.

FISCHER, C. (1982), *To Dwell Among Friends: Personal Networks in Town and City* (Chicago: University of Chicago Press).

FISCHER, P. J. (1988), 'Criminal Activity Among the Homeless: A Study of Arrests in Baltimore', *Hospital and Community Psychiatry*, 39: 46–51.

FLASKERUD, J. H., and UMAN, G. (1993), 'Directions for AIDS Education for Hispanic Women Based on Analysis of Survey Findings', *Public Health Reports*, 108: 298–304.

FOREST, B. (1995), 'West Hollywood as Symbol: The Significance of Place in the Construction of a Gay Identity', *Environment and Planning D: Society and Space*, 13: 133–57.

FOUCAULT, M. (1973), *The Birth of the Clinic: An Archaeology of Medical Perception*, trans. A. M. S. Smith (New York: Pantheon Books).

—— (1977), *Discipline and Punish: The Birth of the Prison*, trans. A. Sheridan, (New York: Vintage Books).

—— (1980), *Power/Knowledge: Selected Interviews and Other Writings, 1972–1977*, ed. C. Gordon, trans. C. Gordon (New York: Pantheon Books).

FRANK, G. (1992), '5 Cities Sued Over Curbs on Homeless', *Los Angeles Times*, 11 September, A3.

FRIEDMAN, S. R., DES JARLAIS, D. C., and STERK, C. E. (1990), 'AIDS and the Social Relations of Intravenous Drug Users', *The Milbank Quarterly*, 68: 85–110.

FUSS, D. (1989), *Essentially Speaking: Feminism, Nature & Difference* (New York: Routledge).

GABER, S. L. (1996), 'From NIMBY to Fair Share: The Development of New

York City's Municipal Shelter Siting Policies, 1980–1990', *Urban Geography*, 17: 294–316.

GALLAGHER, M. L. (1994), 'Homeless—Not Hopeless', *Planning*, 60: 18–21.

Gallup Organization (1995), 'Homeless But Not Hopeless: A Los Angeles Mission Report on What Americans Believe About Homeless People, Their Problems, and Possible Solutions' (Lincoln, Neb.: The Gallup Organization).

GALSTER, G. C., and HILL, E. W. (1992) (eds.), *The Metropolis in Black & White: Place, Power and Polarization* (New Brunswick, NJ: Center for Urban Policy Research, Rutgers University).

GANS, H. J. (1995), *The War Against the Poor: The Underclass and Antipoverty Policy* (New York: Basic Books).

GELBERG, L., LINN, L. S., and LEAKE, B. D. (1988), 'Mental Health, Alcohol and Drug Use, and Criminal History among Homeless Adults', *American Journal of Psychology*, 145: 191–6.

GELTMAKER, T. (1992), 'The Queer Nation Acts Up: Health Care, Politics, and Sexual Diversity in the City of Angels', *Environment and Planning D: Society and Space*, 10: 609–50.

GIDDENS, A. (1984), *The Constitution of Society: Outline of the Theory of Structuration* (Berkeley: University of California Press).

GILBERT, N. (1983), *Capitalism and the Welfare State: Dilemmas of Social Benevolence* (New Haven: Yale University Press).

GILDERBLOOM, J., and APPLEBAUM, R. (1988), *Rethinking Rental Housing* (Philadelphia: Temple University Press).

GILMAN, S. L. (1985), *Difference and Pathology: Stereotypes of Sexuality, Race, and Madness* (Ithaca, NY: Cornell University Press).

—— (1988), *Disease and Representation: Images of Illness from Madness to AIDS* (Ithaca, NY: Cornell University Press).

GILROY, P. (1992), 'The End of Antiracism', in J. Donald and A. Rattansi (eds.), *'Race', Culture and Difference* (London: Sage Publications), 49–61.

GINZBURG, H. M. (1993), 'Federal Response to Needle-Exchange Programs: Part II: A. Science vs. Politics', *Pediatric AIDS and HIV Infection: Fetus to Adolescent*, 4: 88–91.

GOERING, P., WASYLENKI, D., ST ONGE, M., PADUCHAK, D., and LANCEE, W. (1992), 'Gender Differences Among Clients of a Case Management Program for the Homeless', *Hospital and Community Psychiatry*, 43: 160–5.

GOFFMAN, E. (1963), *Stigma: Notes on the Management of Spoiled Identity* (Englewood Cliffs, NJ: Prentice Hall).

GOLDBERG, S. B. (1994), 'Homeless Victory: Santa Ana Law Criminalizes Status', *ABA Journal*, 80: 102–3.

GOLDEN, S. (1992), *The Women Outside: Meanings and Myths of Homelessness* (Berkeley: University of California Press).

GOLDIN, C. S. (1994), 'Stigmatization and AIDS: Critical Issues in Public Health', *Social Science and Medicine*, 39: 1359–66.

GOODMAN, L. A. (1991), 'The Relationship between Social Support and Family Homelessness: A Comparison Study of Homeless and Housed Mothers', *Journal of Community Psychology*, 19: 321–32.

—— DUTTON, M. A., and HARRIS, M. (1995), 'Episodically Homeless Women with

Serious Mental Illness: Prevalence of Physical and Sexual Assault', *American Journal of Orthopsychiatry*, 65: 468–78.

GORDON, L. (1994), 'A New Home for the City's Homeless', *Los Angeles Times*, 15 September, B1.

GOSS, J. (1988), 'The Built Environment and Social Theory: Towards an Architectural Geography', *Professional Geographer*, 40: 392–403.

GOULD, P. (1993), *The Slow Plague: A Geography of the AIDS Pandemic* (Oxford: Blackwell).

GOULD, S. J. (1991), 'The Terrifying Normalcy of AIDS', in N. F. McKenzie (ed.), *The AIDS Reader: Social, Political, and Ethical Issues*, (New York: Meridian/ Penguin), 100–6.

GRANT, J. (1994), *The Drama of Democracy: Contention and Dispute in Community Planning* (Toronto: University of Toronto Press).

GREEN, D. E., McCORMICK, I. A., and WALKEY, F. H. (1987), 'Community Attitudes to Mental Illness in New Zealand Twenty-two Years On', *Social Science and Medicine*, 24: 417–22.

GREGORY, D. (1994), *Geographical Imaginations* (Cambridge, Mass.: Blackwell).

GREGORY, R., SLOVIC, P., and FLYNN, J. (1996), 'Risk Perceptions, Stigma, and Health Policy', *Health & Place*, 4: 213–20.

GUPTA, A., and FERGUSON, J. (1992), 'Beyond "Culture": Space, Identity, and the Politics of Difference', *Cultural Anthropology*, 7: 6–23.

GURZA, A. (1992), 'Homeless Gain Support Against Crackdown', *Orange County Register*, 31 October, B8.

—— (1993*a*), 'Santa Ana is Offered Homeless Settlement', *Orange County Register*, 29 June, B1.

—— (1993*b*), 'Study, Court Hand Homeless Victories', *Orange County Register*, 30 June, B1.

—— (1993*c*), 'Court Stops Trials Against Homeless', *Orange County Register*, 23 July, B1.

—— (1993*d*), 'Santa Ana May Take Another Avenue Against Encampments', *Orange County Register*, 3 December, B7.

—— (1994), 'City Cited for Biased Homeless Policy', *Orange County Register*, 4 February, A2.

—— (1995), 'Ruling Brings Little Hope to Activists', *Orange County Register*, 25 April, A4.

HALL, P. D. (1987), 'A Historical Overview of the Private Nonprofit Sector', in W. W. Powell (ed.), *The Nonprofit Sector: A Research Handbook* (New Haven: Yale University Press), 3–26.

Hamilton, Rabinovitz, and Alschuler, Inc. (1987), 'The Changing Face of Misery: Los Angeles' Skid Row Area in Transition', report prepared for the Community Redevelopment Agency (Los Angeles: Community Redevelopment Agency, City of Los Angeles).

HARRINGTON M. (1981), *The Other America: Poverty in the United States* (New York: Macmillan).

HARRISON, B. and BLUESTONE, B. (1988), *The Great U-Turn: Corporate Restructuring and the Polarizing of America* (New York: Basic Books).

HARTMAN, C. (1986), 'Housing Policies Under the Reagan Administration', in R. G.

Bratt, C. Hartman, and A. Meyerson (eds.), *Critical Perspectives on Housing* (Philadelphia: Temple University Press), 362–77.

HARTZ, D., BANYS, P., and HALL, S. M. (1994), 'Correlates of Homelessness Among Substance Abuse Patients at a VA Medical Center', *Hospital and Community Psychiatry*, 45 5: 491–3.

HARVEY, D. (1973), *Social Justice and the City* (Baltimore: The Johns Hopkins University Press).

—— (1989), *The Condition of Postmodernity: An Enquiry into the Origins of Cultural Change* (Oxford: Blackwell).

—— (1993), 'Class Relations, Social Justice and the Politics of Difference', in M. Keith and S. Pile (eds.), *Place and the Politics of Identity* (London: Routledge), 41–66.

HAWKESWORTH, M. E. (1988), 'Feminist Rhetoric: Discourses on the Male Monopoly of Thought', *Political Theory*, 16: 444–67.

HAYDEN, D. (1984), *Redesigning the American Dream* (New York: W. W. Norton).

—— (1995), *The Power of Place: Urban Landscapes as Public History* (Cambridge, Mass.: MIT Press).

HEIMAN, M. (1990), 'From "Not In My Backyard" to "Not In Anybody's Backyard", Grassroots Challenge to Hazardous Waste Facility Siting', *Journal of the American Planning Association*, 56: 359–62.

HELLINGER, F. J. (1993), 'The Lifetime Cost of Treating a Person with HIV', *Journal of the American Medical Association*, 270 4: 474–8.

HERMAN, D. B., STRUENING, E. L., and BARROW, S. M. (1993), 'Self-Assessed Need for Mental Health Services Among Homeless Adults', *Hospital and Community Psychiatry*, 44 12: 1181–3.

HERNANDEZ, R. (1996*a*), 'Homeless Man Will Challenge City Law', *Orange County Register*, 3 May, B1.

—— (1996*b*), 'Judge Convicts Homeless Man of Violating Anti-Camping Law', *Orange County Register*, 9 May, B5.

—— (1996*c*), 'Is Living on the Streets a Crime?', *Orange County Register*, 13 May, B1.

HEWITT, N. A., and LEBSOCK, S. (1993) (eds.), *Visible Women: New Essays on American Activism* (Urbana, Ill.: University of Illinois Press).

HILL-HOLTZMAN, N. (1992), 'The Battle for Santa Monica's Parks', *Los Angeles Times*, 18 January, B1, B2.

—— (1994), 'Ruling Raises Doubts About Homeless Law', *Los Angeles Times*, 13 February, J3.

HOCH, C. (1991), 'The Spatial Organization of the Urban Homeless: A Case Study of Chicago', *Urban Geography*, 12 2: 137–54.

—— and SLAYTON, R. A. (1989), *New Homeless and Old: Community and the Skid Row Hotel* (Philadelphia: Temple University Press).

HOGAN, R. (1986), 'Community Opposition to Group Homes', *Social Sciences Quarterly*, 67: 442–9.

HOMBS, M. E., and SNYDER, M. (1982), *Homelessness in America: A Forced March to Nowhere* (Washington, DC: Community for Creative Nonviolence).

HOOKS, B. (1990), *Yearnings: Race, Gender and Cultural Politics* (Boston, South End Press).

HOPPER, K. (1990), 'Public Shelter as "a Hybrid Institution": Homeless Men in Historical Perspective', *Journal of Social Issues*, 46: 13–29.

—— and BAUMOHL, J. (1994), 'Held in Abeyance: Rethinking Homelessness and Advocacy', *American Behavioral Scientist*, 37: 522–52.

—— and COX, L. S. (1986), 'Litigation in Advocacy for the Homeless', in J. Erickson and C. Wilhelm (eds.), *Housing the Homeless* (New Brunswick, NJ: Center for Urban Policy Research), 303–14.

—— SUSSER, E., and CONOVER, S. (1985), 'Economies of Makeshift: Deindustrialization and Homelessness in New York City', *Urban Anthropology*, 14: 183–236.

HU, D. J., FLEMING, P. L., CASTRO, K. G., JONES, J. L., BUSH, T. J., HANSON, D., CHU, S. Y., KAPLAN, J., and WARD, J. W. (1995), 'How Important Is Race/Ethnicity as an Indicator of Risk for Specific AIDS-Defining Conditions?', *Journal of Acquired Immune Deficiency Syndromes and Human Retrovirology*, 10: 374–80.

HUBBARD, P. (forthcoming), 'Red-Light Districts and Toleration Zones: Geographies of Female Street Prostitution in England and Wales', *Area*, 1–13.

HURLEY, A. (1995), *Environmental Inequalities: Class, Race and Industrial Pollution in Gary, Indiana, 1945–1980* (Chapel Hill, NC: University of North Carolina Press).

HUTSON, S., and LIDDIARD, M. (1994), *Youth Homelessness: The Construction of a Social Issue* (London: Macmillan).

Institute of Health Policy Studies (1993), 'HIV Prevention in California: Final Report HIV Education and Prevention Evaluation' (Sacramento, Calif.: California Department of Health Services, Office of AIDS).

Institute of Medicine, Committee on Health Care for Homeless People (1988), *Homelessness, Health, and Human Needs* (Washington, DC: National Academy Press).

JACKSON, P. (1987) (ed.), *Race and Racism: Essays in Social Geography* (London: Allen & Unwin).

—— (1989), *Maps of Meaning: An Introduction to Cultural Geography* (London: Unwin Hyman).

—— (1991), 'The Crisis of Representation and the Politics of Position', *Environment and Planning D: Society and Space*, 9: 131–4.

—— (1993), 'Visibility and Voice' (Editorial), *Environment and Planning D: Society and Space*, 11: 123–6.

—— and PENROSE, J. (1993) (eds.), *Constructions of Race, Place and Nation* (London: UCL Press).

JONES, E. E., FARINA, A., HASTORF, A. H., MARKUS, H., MILLER, D. T., SCOTT, R. A., and FRENCH, R. S. (1984), *Social Stigma: The Psychology of Marked Relationships* (New York: W. H. Freeman).

KAPLAN, E. (1993), 'Federal Response to Needle-Exchange Programs: Part II: B. Needle-Exchange Research: The New Haven Experience', *Pediatric AIDS and HIV Infection: Fetus to Adolescent*, 4: 92–6.

KATZ, I. (1981), *Stigma: A Social Psychological Analysis* (Hillsdale, NJ: Lawrence Erlbaum Associates).

KATZ, M. (1989), *The Undeserving Poor: From the War on Poverty to the War on Welfare* (New York: Pantheon).

KAZMIN, A. L. (1992), 'Marchers Rally Against Rash of Gay-Bashing Incidents in Silver Lake', *Los Angeles Times*, 14 November, B3.

KEARNS, R. A. (1992), 'The Stress of Incipient Homelessness', *Housing Studies*, 7: 274–87.

—— (1996), 'AIDS and Medical Geography: Embracing the Other?', *Progress in Human Geography*, 20: 123–31.

—— and SMITH, C. J. (1993), 'Housing Stressors and Mental Health Among Marginalized Urban Populations', *Area*, 25: 267–78.

KEIGHER, S. M. (1991), (ed.), *Housing Risks and Homelessness Among the Urban Elderly* (New York: Haworth Press).

KELLER, L. R., and SARIN, R. K. (1988), 'Equity in Social Risk: Some Empirical Observations', *Risk Analysis*, 8: 135–46.

—— —— (1995), 'Fair Processes for Societal Decisions Involving Distributional Inequalities', *Risk Analysis*, 13: 49–59.

KELLY, J., ST LAWRENCE, J. S., STEVENSON, Y. L., HAUTH, A. C., KALICHMAN, S. C., DIAZ, Y. E., BRASFIELD, T. L., KOOB, J. J., and MORGAN, M. G. (1992), 'Community AIDS/HIV Risk Reduction: The Effects of Endorsements by Popular People in Three Cities', *American Journal of Public Health*, 82: 1483–9.

KIRBY, A. (1982), *The Politics of Location: An Introduction* (London: Methuen).

KIRSCHENMAN, J., and NECKERMAN, K. M. (1991), '"We'd Love to Hire Them, But . . .": The Meaning of Race for Employers', in C. Jencks and P. E. Peterson (eds.), *The Urban Underclass* (Washington, DC: The Brookings Institution), 203–34.

KLINE, A., KLINE, E., and OKEN, E. (1992), 'Minority Women and Sexual Choice in the Age of AIDS', *Social Science and Medicine*, 34: 447–57.

KODRAS, J. E., and JONES, J. P., III (1990) (eds.), *Geographic Dimensions of United States Social Policy* (London: Edward Arnold).

KONDRATAS, A. (1991), 'Ending Homelessness: Policy Challenges', *American Psychologist*, 46 11: 1226–32.

KOZOL, J. (1988), *Rachel and Her Children: Homeless Families in America* (New York: Fawcett Columbine).

KRESS, J. B. (1994), 'Homeless Fatigue Syndrome: The Backlash Against the Crime of Homelessness in the 1990s', *Social Justice*, 21: 85–108.

LA GORY, M., RITCHEY, F. J., and MULLIS, J. (1990), 'Depression among the Homeless', *Journal of Health and Social Behavior*, 31: 87–101.

LAKE, R. W. (1990), 'Urban Fortunes: The Political Economy of Place: A Commentary', *Urban Geography*, 11: 179–84.

—— (1993), 'Rethinking NIMBY', *Journal of the American Planning Association*, 59: 87–93.

—— (1994), 'Negotiating Local Autonomy', *Political Geography*, 13: 423–42.

—— and DISCH, L. (1992), 'Structural Constraints and Pluralist Contradictions in Hazardous Waste Regulation', *Environment and Planning A*, 24: 663–81.

—— and JOHNS, R. A. (1990), 'Legitimation Conflicts: The Politics of Hazardous Waste Siting Law', *Urban Geography*, 11: 488–508.

LAUBER, D. (1973), 'Recent Cases in Exclusionary Zoning', report prepared for the American Society of Planning Officials (Chicago: American Society of Planning Officials).

—— (1985), 'Mainstreaming Group Homes', *Planning*, 14–18.

—— with BANKS, F. S., Jr. (1974), 'Zoning for Family and Group Care Facilities', report prepared for the American Society of Planning Officials (Chicago: American Society of Planning Officials).

LAW, R. (1991), '"Not In My City": Municipal Responses to Homelessness', Working paper, Los Angeles Homelessness Project (University of Southern California).

—— (1996), '"Not In My City': Local Government Planners and Homelessness', Working paper, Department of Geography, University of Otago (Dunedin, New Zealand).

—— and WOLCH, J. R. (1991), 'Homelessness and Economic Restructuring', *Urban Geography*, 12 2: 105–36.

—— —— and TAKAHASHI, L. (1993), 'Defense-less Territory: Workers, Communities and the Decline of Military Production in Los Angeles', *Environment and Planning C: Government and Policy*, 11: 291–315.

LAWS, G. (1992), 'Emergency Shelter Networks in an Urban Area: Serving the Homeless in Metropolitan Toronto', *Urban Geography*, 13: 99–126.

—— (1994), 'Aging, Contested Meanings, and the Built Environment', *Environment and Planning A*, 26: 1787–1802.

—— (1995), 'Embodiment and Emplacement: Identities, Representation and Landscape in Sun City Retirement Communities', *International Journal of Aging and Human Development*, 40: 253–80.

LAYTON, M. C., CANTWELL, M. F., DORSINVILLE, G. J., VALWAY, S. E., ONORATO, I. M., and FRIEDEN, T. R. (1995), 'Tuberculosis Screening among Homeless Persons with AIDS Living in Single-Room-Occupancy Hotels', *American Journal of Public Health*, 85: 1555–9.

LEBOW, J. M., O'CONNELL, J. J., ODDLEIFSON, S., GALLAGHER, K. M., SEAGE, G. R., III, and FREEDBERG, K. A. (1995), 'AIDS Among the Homeless of Boston: A Cohort Study', *Journal of Acquired Immune Deficiency Syndromes and Human Retrovirology*, 8: 292–6.

LEE, B. A., JONES, S. H., and LEWIS, D. W. (1990), 'Public Beliefs About the Causes of Homelessness', *Social Forces*, 69: 253–65.

—— LEWIS, D. W., and JONES, S. H. (1992), 'Are the Homeless to Blame? A Test of Two Theories', *The Sociological Quarterly*, 33: 535–52.

LEE, D. (1995), 'Crisis of Confidence: Orange County Survey Reflects Toll of Bankruptcy', *Los Angeles Times*, 19 September, D1.

LEFEBVRE, H. (1991), *The Production of Space*, trans. D. Nicholson-Smith (Oxford: Blackwell).

LEMERT, E. M. (1951), *Social Pathology: A Systematic Approach to the Theory of Sociopathic Behaviours* (New York: McGraw-Hill).

—— (1972), *Human Deviance, Social Problems, and Social Control*, second edition (Englewood Cliffs, NJ: Prentice-Hall).

LEO, J. (1994), 'The (Local) Politics of Housing', *U.S. News & World Report*, 29 August/5 September, 20.

LESHER, D. (1996), 'Welfare Payment Reductions a Step Nearer', *Los Angeles Times*, 10 October, A3, A31.

LIEBOW, E. (1993), *Tell Them Who I Am: The Lives of Homeless Women* (New York: The Free Press).

LIGGETT, H., and PERRY, D. C. (1995) (eds.), *Spatial Practices: Critical Explorations in Social/Spatial Theory* (Thousand Oaks, Calif.: Sage Publications).

LINK, B. G., SUSSER, E., STUEVE, A., and PHELAN, J. (1994), 'Lifetime and Five-year Prevalence of Homelessness in the United States', *American Journal of Public Health*, 84 12: 1907–12.

LOGAN, J. R., and MOLOTCH, H. L. (1987), *Urban Fortunes: The Political Economy of Place* (Berkeley: University of California Press).

LOW, N. P., and GLEESON, B. J. (1997), 'Justice in and to the Environment: Ethical Uncertainties and Political Practices', *Environment and Planning A*, 29: 21–42.

LYNCH, K. (1960), *The Image of the City* (Cambridge, Mass.: MIT Press).

McCHESNEY, K. Y. (1990), 'Family Homelessness: A Systemic Problem', *Journal of Social Issues*, 46: 191–205.

MACDONALD, H. (1994), 'Free Housing Yes, Free Speech No', *The Wall Street Journal*, 8 August, 1.

McKEGANEY, N. P. (1994), 'Prostitution and HIV: What Do We Know and Where Might Research Be Targeted in the Future?', *AIDS*, 8: 1215–26.

MACLACHLAN, J. (1992), 'Managing AIDS: A Phenomenology of Experiment, Empowerment and Expediency', *Critique of Anthropology*, 12: 433–56.

MACWILLIAMS, B. (1993), 'Judge Upholds Limit on Homeless Policy', *Orange County Register*, 30 October, B1.

—— (1995a), 'County Urged to Drop Cases vs. Homeless', *Orange County Register*, 22 June, B4.

—— (1995b), 'Deal Lets Homeless Avoid Trial', *Orange County Register*, 23 June, B7.

MAGAÑA, R., and CARRIER, J. (1991), 'Mexican and Mexican American Male Sexual Behavior and Spread of AIDS in California', *The Journal of Sex Research*, 28: 425–41.

MAIR, A. (1986), 'The Homeless and the Post-Industrial City', *Political Geography*, 5: 351–68.

MARCUSE, P. (1988), 'Perspectives on Homelessness', *Urban Affairs Quarterly*, 23 4: 647–56.

MARIN, G. (1989), 'AIDS Prevention Among Hispanics: Needs, Risk Behaviors, and Cultural Values', *Public Health Reports*, 104: 411–15.

—— (1993), 'Defining Culturally Appropriate Community Interventions: Hispanics as a Case Study', *Journal of Community Psychology*, 21: 149–61.

MARKOS, A. R., WADE, A. A. H., and WALZMAN, M. (1994), 'The Adolescent Male Prostitute and Sexually Transmitted Diseases, HIV and AIDS', *Journal of Adolescence*, 17: 123–30.

MARSHALL, J. (1991), 'Women and Homelessness', *Journal of Maternal and Child Health*, 372–4.

MARTINEZ, G. (1991a), 'Lawsuit Settled; Homeless to Get $400,000', *Los Angeles Times*, 21 August, A21.

—— (1991b), 'Windfall for the Homeless Fails to Offer Better Life', *Los Angeles Times*, 2 December, A3, A21.

MASSEY, D. (1994), *Space, Place and Gender* (Cambridge: Polity Press).

MAYS, V., and COCHRAN, S. (1988), 'Issues in the Perception of AIDS Risk and Risk Reduction Activities by Black and Hispanic/Latina Women', *American Psychologist*, 41: 949–57.

MECHANIC, D., and ROCHEFORT, D. A. (1990), 'Deinstitutionalization: An Appraisal of Reform', *Annual Review of Sociology*, 16: 301–27.

MERRIFIELD, A., and SWYNGEDOUW, E. (1997) (eds.), *The Urbanization of Injustice* (Washington Square, NY: University Press).

MEYER, W. B. (1995), 'NIMBY Then and Now: Land-Use Conflict in Worcester, Massachusetts, 1876–1900', *Professional Geographer*, 47: 298–308.

—— and BROWN, M. (1989), 'Locational Conflict in a Nineteenth-Century City', *Political Geography Quarterly*, 8: 107–22.

MILLER, J. (1994), 'Camping Ban To Go To State Supreme Court', *Orange County Register*, 13 May, B5.

MINOW, M. (1990), *Making All the Difference: Inclusion, Exclusion, and American Law* (Ithaca, NY: Cornell University Press).

MISHRA, S. I. (1994), 'Health Care Needs in Orange County, California: An Assessment of Sheltered Homeless Persons', report prepared for the United Way of Orange County Health Care Council (Irvine, Calif.: Center for Health Policy and Research, University of California-Irvine).

MITCHELL, D. (1995), 'The End of Public Space? People's Park, Definitions of the Public, and Democracy', *Annals of the Association of American Geographers*, 85: 108–33.

MOEHRINGER, J. R. (1994), 'Homeless Aid Is Casualty of Fiscal Crisis', *Los Angeles Times*, 28 December, A1, A12.

MOLLENKOPF, J. H. (1983), *The Contested City* (Princeton: Princeton University Press).

MORALES, J., and BOK, M. (1992) (eds.), *Multicultural Human Services for AIDS Treatment and Prevention: Policy, Perspectives, and Planning* (New York: The Haworth Press).

MORSE, G. A., CALSYN, R. J., ALLEN, G., TEMPELHOFF, B., and SMITH, R. (1992), 'Experimental Comparison of the Effects of Three Treatment Programs for Homeless Mentally Ill People', *Hospital and Community Psychiatry*, 43: 1005–10.

NAMASTE, K. (1996), 'Genderbashing: Sexuality, Gender, and the Regulation of Public Space', *Environment and Planning D: Society and Space*, 14: 221–40.

NEWTON, J. (1997), 'L.A. Homeless Don't Share in Rewards of Recovery', *Los Angeles Times*, 19 May, A14–A15.

NI, C. C. (1994), 'Camp Curbside', *Los Angeles Times*, 26 October, B1, B8.

NICHOLSON, L., and SEIDMAN, S. (1995) (eds.), *Social Postmodernism: Beyond Identity Politics* (Cambridge: Cambridge University Press).

NOBLE, G. R., PARRA, W. C., and HOLMAN, P. B. (1991), 'Organizational Structure and Resources of CDC's HIV-AIDS Prevention Program', *Public Health Reports*, 106: 604–7.

NYAMATHI, A. (1992), 'Comparative Study of Factors Relating to HIV Risk Level of Black Homeless Women', *Journal of Acquired Immune Deficiency Syndrome*, 5: 222–8.

OBOLER, S. (1995), *Ethnic Labels, Latino Lives: Identity and the Politics of (Re)Presentation in the United States* (Minneapolis: University of Minnesota Press).

O'HARE, M., and SANDERSON, D. (1993), 'Facility Siting and Compensation: Lessons from the Massachusetts Experience', *Journal of Public Policy Analysis and Management*, 12: 364–76.

OMI, M., and WINANT, H. (1994), *Racial Formation in the United States: From the 1960s to the 1980s* (New York: Routledge and Kegan Paul).

Orange County HIV Planning Advisory Council (1996), 'County of Orange Ryan White Title I Application 1996' (Santa Ana, Calif.: Orange County Health Care Agency).

Orange County HIV Prevention Planning Committee (1995), 'Plan for the Prevention of HIV in Orange County, California 1996–1999' (Santa Ana, Calif.: Orange County Health Care Agency).

OSMOND, M. W., WAMBACH, K. G., HARRISON, D. F., BYERS, J., LEVINE, P., IMERSHEIN, A., and QUADAGNO, D. (1993), 'The Multiple Jeopardy of Race, Class, and Gender for AIDS Risk Among Women', *Gender & Society*, 7: 99–120.

PATTON, C. (1990), *Inventing AIDS* (New York, Routledge).

—— (1994), *Last Served? Gendering the HIV Pandemic* (London: Taylor & Francis).

PEATTIE, L. (1968), 'Reflections on Advocacy Planning', *Journal of the American Institute of Planners*, 34: 80–8.

PHILLIPS, K. (1990), *The Politics of Rich and Poor: Wealth and the American Electorate in the Reagan Aftermath* (New York: Harper Collins).

PIVEN, F. F., and CLOWARD, R. (1971), *Regulating the Poor: The Functions of Public Welfare* (New York: Pantheon).

—— —— (1977), *Poor People's Movements: Why They Succeed, How They Fail* (New York: Pantheon).

—— —— (1982), *The New Class War: Reagan's Attack on the Welfare State and Its Consequences* (New York: Pantheon).

PLOTKIN, S. (1987), *Keep Out: The Struggle for Land Use Control* (Berkeley: University of California Press).

POPPER, F. J. (1981), *The Politics of Land-Use Reform* (Madison, Wis.: University of Wisconsin Press).

PRED, A. (1985), 'The Social Becomes the Spatial, the Spatial Becomes the Social: Enclosures, Social Change and the Becoming of Places in Skane', in J. Urry and D. Gregory (eds.), *Social Relations and Spatial Structures* (London: Macmillan), 337–65.

PULIDO, L. (1994), 'Restructuring and the Contraction and Expansion of Environmental Rights in the United States', *Environment and Planning A*, 26: 915–36.

—— (1996a), *Environmentalism and Economic Justice: Two Chicano Struggles in the Southwest* (Tucson, Ariz.: University of Arizona Press).

—— (1996b), 'A Critical Review of the Methodology of Environmental Racism Research', *Antipode*, 28: 142–59.

—— (1998), 'Community, Place, and Identity,' in J. P. Jones, H. Nastand, and S. Roberts (eds.), *Thresholds in Feminist Geography* (Lanham, Maryland: Rowman & Littlefield), 11–28.

QUIMBY, E. (1992), 'Anthropological Witnessing for African Americans: Power, Responsibility, and Choice in the Age of AIDS', in G. Herdt and S. Lindenbaum (eds.), *The Time of AIDS: Social Analysis, Theory, and Method* (Newbury Park, Calif.: Sage Publications), 159–84.

—— (1993), 'Obstacles to Reducing AIDS Among African Americans', *Journal of Black Psychology*, 19: 215–22.

—— and FRIEDMAN, S. R. (1989), 'Dynamics of Black Mobilization against AIDS in New York City', *Social Problems*, 36: 403–15.

RAHIMIAN, A., WOLCH, J., and KOEGEL, P. (1992), 'A Model of Homeless Migration:

Homeless Men in Skid Row, Los Angeles', *Environment and Planning A*, 24: 1317–36.

RELPH, E. (1976), *Place and Placelessness* (London: Pion).

RICH, A. (1986), *Blood, Bread, and Poetry: Selected Prose 1979–1985* (New York: W. W. Norton).

RICHARDSON, L. (1996), 'Feeding Their Fear: Food Stamp Changes Leave Some Recipients Expecting the Worst', *Los Angeles Times*, 2 October, B1, B10.

RINGHEIM, K. (1990), *At Risk of Homelessness: The Roles of Income and Rent* (New York: Praeger).

RITCHEY, F. J., LA GORY, M., and MULLIS, J. (1991), 'Gender Differences in Health Risks and Physical Symptoms Among the Homeless', *Journal of Health and Social Behavior*, 32: 33–48.

RIVERA, C. (1996), 'Welfare Law's Job Goal May Be Impossible', *Los Angeles Times*, 4 November, A1, A24–A25.

ROBERTSON, M. J. (1991), 'Homeless Women with Children', *American Psychologist*, 46: 1198–204.

—— and GREENBLATT, M. (1992) (eds.), *Homelessness: A National Perspective* (New York: Plenum Press).

ROCHA, E., and DEAR, M. J. (1989), 'Gaining Acceptance for Community-Based Service Facilities: An Annotated Bibliography', Working Paper, Department of Geography (Los Angeles: University of Southern California).

ROGERS, A., and UTO, R. (1987), 'Residential Segregation Retheorized: A View from Southern California', in P. Jackson (ed.), *Race and Racism: Essays in Social Geography*, (London: Allen & Unwin), 50–73.

ROGERS, T. F., SINGER, E., and IMPERIO, J. (1993), 'Poll Trends: AIDS—An Update', *Public Opinion Quarterly*, 57: 92–114.

ROMÁN, D. (1993), '*Fierce Love* and Fierce Response: Intervening in the Cultural Politics of Race, Sexuality, and AIDS', *Journal of Homosexuality*, 26: 195–219.

ROMNEY, L. (1994), 'High Court Showdown on Homeless Issue Nears', *Los Angeles Times*, 6 December, A1, A25.

ROPER, W. L. (1991), 'Current Approaches to Prevention of HIV Infections', *Public Health Reports*, 106: 111–15.

ROPERS, R. H. (1988), *The Invisible Homeless: A New Urban Ecology* (New York: Insight Books).

ROSE, G. (1993), *Feminism and Geography: The Limits of Geographical Knowledge* (Cambridge: Polity Press).

ROSE, J. B. (1993), 'A Critical Assessment of New York City's Fair Share Criteria', *Journal of the American Planning Association*, 59: 97–100.

ROSSI, P. (1989), *Down and Out in America: The Origins of Homelessness* (Chicago: University of Chicago Press).

ROWE, S. (1997), personal conversation (Los Angeles).

—— and WOLCH, J. (1990), 'Social Networks in Time and Space: Homeless Women in Skid Row, Los Angeles', *Annals of the Association of American Geographers*, 80: 184–204.

RUDDICK, S. M. (1996), *Young and Homeless in Hollywood: Mapping Social Identities* (New York: Routledge).

RUDZINSKI, K. A., MARCONI, K. M., and McKINNEY, M. M. (1994), 'Federal

Funding For Health Services Research On AIDS, 1986–1991', *Health Affairs*, 13: 261–6.

RUSHING, W. A. (1969) (ed.), *Deviant Behavior and Social Process* (Chicago: Rand McNally).

—— (1995), *The AIDS Epidemic: Social Dimensions of an Infectious Disease* (Boulder, Colo.: Westview Press).

RYAN, M. (1989), *Politics and Culture: Working Hypotheses for a Post Revolutionary Society* (Baltimore: The Johns Hopkins University Press).

SACHS, A. (1993), 'A Right to Sleep Outside? Laws Designed to Keep Homeless Out of Public Areas Spur Lawsuits', *ABA Journal*, 79: 38.

SAGARIN, E. (1975), *Deviants and Deviance: An Introduction to the Study of Disvalued People and Behavior* (New York: Praeger).

SAHAGUN, L. (1987a), '"River-Bottom" People: Dirt, Debate, Dilemma', *Los Angeles Times*, 6 October, 3, 16.

—— (1987b), 'Riverside Police Raid Homeless Camp', *Los Angeles Times*, 10 October, 27.

—— (1990), 'L.A. River Offers Refuge for Homeless Immigrants', *Los Angeles Times*, 13 August, A1, A3.

SANDERCOCK, L., and FORSYTH, A. (1992), 'A Gender Agenda: New Directions for Planning Theory', *Journal of the American Planning Association*, 58: 49–59.

SANTOYO, G. (1993a), 'Advocates of Homeless Fight New Law', *Orange County Register*, 22 December, B4.

—— (1993b), 'Controlling Camping', *Orange County Register*, 30 December, 1.

SCHIETINGER, H., COBURN, J., and LEVI, J. (1995), 'Community Planning for HIV Prevention: Findings from the First Year', *AIDS and Public Policy Journal*, 10: 140–7.

SCHILLER, N. G., CRYSTAL, S., and LEWELLEN, D. (1994), 'Risky Business: The Cultural Construction of AIDS Risk Groups', *Social Science and Medicine*, 38: 1337–46.

SCHUR, E. M. (1971), *Labeling Deviant Behavior: Its Sociological Implications* (New York: Harper & Row).

—— (1980), *The Politics of Deviance: Stigma Contests and the Uses of Power* (Englewood Cliffs, NJ: Prentice-Hall).

SCHWARTZ, B., and KURTZMAN, L. (1988), 'Tough New Homeless Policy Triggers Furor in Santa Ana', *Los Angeles Times*, 10 July, I3, I32.

SCHWARTZ, D. C., DEVANCE-MANZINI, D., and FAGAN, T. (1991), 'Preventing Homelessness: A Study of State and Local Homelessness Prevention Programs', prepared for the American Affordable Housing Institute (New Brunswick, NJ: American Affordable Housing Institute, Rutgers University).

SCOTT, A. J. (1986), 'High Technology Industry and Territorial Development: The Rise of the Orange County Complex, 1955–1984', *Urban Geography*, 7: 3–45.

—— (1993), *Technopolis: High Technology Industry and Regional Development in Southern California* (Berkeley: University of California Press).

—— and SOJA, E. W. (1996) (eds.), *The City: Los Angeles and Urban Theory at the End of the Twentieth Century* (Berkeley: University of California Press).

SCOTT, R. A. (1972), 'A Proposed Framework for Analyzing Deviance as a Property of Social Order', in R. A. Scott and J. D. Douglas (eds.), *Theoretical Perspectives on Deviance* (New York: Basic Books), 9–36.

Scott, R. A., and Douglas, J. D. (1972) (eds.), *Theoretical Perspectives on Deviance* (New York: Basic Books).

Segal, S. P., Baumohl, J., and Moyles, E. W. (1980), 'Neighborhood Types and Community Reaction to the Mentally Ill: A Paradox of Intensity', *Journal of Health and Social Behavior*, 21: 345–59.

Selik, R. M., Chu, S. Y., and Buehler, J. W. (1993), 'HIV Infection as Leading Cause of Death Among Young Adults in US Cities and States', *Journal of the American Medical Association*, 269: 2991–4.

Shaffer, G. (1992), 'Santa Ana Homeless Put on Notice', *Orange County Register*, 3 September, B1.

Shannon, G. W., Pyle, G. F. and Bashshur R. L. (1991), *The Geography of AIDS: Origins and Course of an Epidemic* (New York: The Guilford Press).

Shaw, N. S. (1991), 'Preventing AIDS Among Women: The Role of Community Organizing', in N. F. McKenzie (ed.), *The AIDS Reader: Social, Political, and Ethical Issues* (New York: Meridian/Penguin), 505–21.

Shields, R. (1991), *Places on the Margin: Alternative Geographies of Modernity* (London: Routledge).

Shinn, M. and Gillespie, C. (1994), 'The Roles of Housing and Poverty in the Origins of Homelessness', *American Behavioral Scientist*, 37 4: 505–21.

—— Knickman, J. R., and Weitzman, B. C. (1991), 'Social Relationships and Vulnerability to Becoming Homeless Among Poor Families', *American Journal of Community Psychology*, 46: 1180–7.

Sibley, D. (1995), *Geographies of Exclusion: Society and Difference in the West* (London: Routledge).

Silver, C. (1985), 'Neighborhood Planning in Historical Perspective', *Journal of the American Planning Association*, 51: 161–74.

Singer, M., Flores, C., Davison, L., Burke, G., Castillo, Z., Scanlon, K., and Rivera, M. (1990a), 'SIDA: The Economic, Social, and Cultural Context of AIDS among Latinos', *Medical Anthropology Quarterly*, 4: 72–114.

—— Castillo, Z., Davison, L., and Flores, C. (1990b), 'Owning AIDS: Latino Organizations and the AIDS Epidemic', *Hispanic Journal of Behavioral Sciences*, 12: 196–211.

—— Flores, C., Davison, L., Burke, G., and Castillo, Z. (1991), 'Puerto Rican Community Mobilizing in Response to the AIDS Crisis', *Human Organization*, 50: 73–81.

Sinnock, P., Murphy, P. E., Baker, T. G., and Bates, R. (1991), 'First Three Years of the National AIDS Clearinghouse', *Public Health Reports*, 106: 634–56.

Small, S. (1994), *Racialised Barriers: The Black Experience in the United States and England in the 1980s* (London: Routledge).

Smith, C. A., Smith, C. J., Kearns, R. A., and Abbott, M. W. (1993), 'Housing Stressors, Social Support and Psychological Distress', *Social Science and Medicine*, 37: 603–12.

Smith, C. J. (1981), 'Residential Proximity and Community Acceptance of the Mentally Ill', *Journal of Operational Psychiatry*, 12: 2–12.

—— (1989), 'Privatization and the Delivery of Mental Health Services', *Urban Geography*, 10 2: 186–95.

—— and Hanham, R. Q. (1981), 'Any Place But Here! Mental Health Facilities as Noxious Neighbors', *Professional Geographer*, 33: 326–34.

SMITH, D. M. (1979), *Where the Grass is Greener: Living in an Unequal World* (Harmondsworth: Penguin Books).

—— *Geography and Social Justice* (Oxford: Blackwell).

SMITH, J. (1996), 'Arresting the Homeless for Sleeping in Public: A Paradigm for Expanding the Robinson Doctrine', *Columbia Journal of Law and Social Problems*, 29: 293–335.

SNOW, D. A., and ANDERSON, L. (1993), *Down on Their Luck: A Study of Homeless Street People* (Berkeley: University of California Press).

SOJA, E. W. (1989), *Postmodern Geographies: The Reassertion of Space in Critical Social Theory* (London: Verso).

—— (1996), *Thirdspace: Journeys to Los Angeles and Other Real-and-Imagined Places* (Cambridge, Mass.: Blackwell).

—— and HOOPER, B. (1993), 'The Spaces that Difference Makes: Some Notes on the Geographical Margins of the New Cultural Politics', in M. Keith and S. Pile (eds.), *Place and the Politics of Identity* (London: Routledge), 183–205.

SOLARZ, A., and BOGAT, G. A. (1990), 'When Social Support Fails: The Homeless', *Journal of Community Psychology*, 18: 79–96.

SOLOMON, P. (1983), 'Analyzing Opposition to Community Residential Facilities for Troubled Adolescents', *Child Welfare*, 62: 361–6.

SONTAG, S. (1988), *AIDS and Its Metaphors* (New York: Farrar, Straus and Giroux).

SPITZER, S. (1975), 'Toward a Marxian Theory of Deviance', *Social Problems*, 22: 638–51.

—— and DENZIN, N. K. (1968) (eds.), *The Mental Patient: Studies in the Sociology of Deviance* (New York: McGraw-Hill).

STALLYBRASS, P., and WHITE, A. (1986), *The Politics and Poetics of Transgression* (Ithaca, NY: Cornell University Press).

STARK, L. R. (1994), 'The Shelter as "Total Institution"', *American Behavioral Scientist*, 37: 553–62.

STONER, M. R. (1989), *Inventing a Non-Homeless Future: A Public Policy Agenda for Preventing Homelessness* (New York: Peter Lang).

—— (1995), *The Civil Rights of Homeless People: Law, Social Policy, and Social Work Practice* (New York: Aldine de Gruyter).

SUSSER, E. S., LIN, S. P., CONOVER, S. A., and STRUENING, E. L. (1991), 'Childhood Antecedents to Homelessness in Psychiatric Patients', *American Journal of Psychiatry*, 148: 1026–30.

SUSSER, I. (1996), 'The Construction of Poverty and Homelessness in US Cities', *Annual Review of Anthropology*, 25: 411–35.

SUSSKIND, L., and CRUIKSHANK, J. (1987), *Breaking the Impasse: Consensual Approaches to Resolving Public Disputes* (New York: Basic Books).

TAKAHASHI, L. M. (1996), 'A Decade of Understanding Homelessness: From Characterization to Representation', *Progress in Human Geography*, 29: 291–310.

—— (1997a), 'The Socio-Spatial Stigmatization of Homelessness and HIV/AIDS: Toward an Explanation of the NIMBY Syndrome', *Social Science and Medicine*, 45: 903–14.

—— (1997b), 'When Does Race Matter? Exploring Socio-Demographic Variation in Attitudes Toward Controversial Facilities' (Research Note), *Urban Geography*, 18: 451–9.

—— (1998), 'Information and Attitudes Toward Mental Health Care Facilities:

Implications for Addressing the NIMBY Syndrome', *Journal of Planning Education and Research*, 17: 119–30.

—— and DEAR, M. J. (1997), 'The Changing Dynamics of Community Attitudes Toward Human Services', *Journal of the American Planning Association*, 63: 79–93.

—— and SMUTNY, G. (forthcoming), 'Community Planning for HIV/AIDS Prevention in Orange County, California', *Journal of the American Planning Association*.

—— and WOLCH, J. R. (1994), 'Differences in Health and Welfare Between Homeless and Homed Welfare Applicants in Los Angeles County', *Social Science and Medicine*, 38: 1401–13.

TAYLOR, S. M., and DEAR, M. J. (1981), 'Scaling Community Attitudes Toward the Mentally Ill', *Schizophrenia Bulletin*, 7: 225–40.

—— ELLIOTT, S. J., and KEARNS, R. A. (1988), 'The Housing Experience of Chronically Mentally Disabled Clients in Hamilton, Ontario', *Canadian Geographer*, 33: 58–76.

THRIFT, N. J. (1996), *Spatial Formations* (London: Sage).

TIEBOUT, C. M. (1956), 'A Pure Theory of Local Expenditures', *Journal of Political Economy*, 64: 416–24.

'TOM' (1996), personal conversation (Los Angeles).

TORO, P. A., and MCDONELL, D. M. (1992), 'Beliefs, Attitudes, and Knowledge About Homelessness: A Survey of the General Public', *American Journal of Community Psychology*, 20: 53–80.

TORREY, E. F. (1988), *Nowhere to Go: The Tragic Odyssey of the Homeless Mentally Ill* (New York: Harper & Row).

TOULMIN, S. (1989), *Cosmopolis: The Hidden Agenda of Modernity* (New York: Free Press).

TRAUB, S. H., and LITTLE, C. B. (1985) (eds.), *Theories of Deviance* (Itasca, Ill.: F.E. Peacock).

TREICHLER, P. (1988), 'AIDS, Homophobia, and Biomedical Discourse: An Epidemic of Signification', in D. Crimp (ed.), *AIDS: Cultural Analysis, Cultural Activism* (Cambridge, Mass.: MIT Press), 31–70.

TRINGO, J. L. (1970), 'The Hierarchy of Preference Toward Disability Groups', *Journal of Special Education*, 4: 295–306.

TUAN, Y. F. (1979), *Landscapes of Fear* (New York: Pantheon Books).

TUCKER, W. (1991), 'How Housing Regulations Cause Homelessness', *Public Interest*, 102: 78–88.

US Department of Housing and Urban Development (HUD) (1984), 'The Extent of Homelessness in America: A Report to the Secretary on the Homeless and Emergency Shelters' (Washington, DC: US Department of Housing and Urban Development).

US House of Representatives (1992), 'The Politics of AIDS Prevention: Science Takes a Time Out', Report 102–1047 (Washington, DC: 102nd Congress, 2nd session).

VALDISERRI, R. O., AULTMAN, T. V., and CURRAN, J. W. (1995), 'Community Planning: A National Strategy to Improve HIV Prevention Programs', *Journal of Community Health*, 20: 87–100.

VALENTINE, G. (1993), '(Hetero)sexing Space: Lesbian Perceptions and Experiences of Everyday Spaces', *Environment and Planning D: Society and Space*, 9: 395–413.

VALETTA, W. (1993), 'Siting Public Facilities on a Fair Share Basis in New York City', *Urban Lawyer*, 25: 1–20.

VANDER KOOI, R. (1973), 'The Main Stem: Skid Row Revisited', *Society*, 10: 64–71.

VANDERPRIEM, D. (1996), 'AIDS Hospice—Local Government Teams with Non-profit Organizations', *California Planner*, 3: 7.

VAN VUGT, J. P. (1994) (ed.), *AIDS Prevention and Services: Community Based Research* (Westport, Conn: Bergin & Garvey).

VENESS, A. R. (1993), 'Neither Homed Nor Homeless: Contested Definitions and the Personal Worlds of the Poor', *Political Geography*, 12: 319–40.

—— (1994), 'Designer Shelters as Models and Makers of Home: New Responses to Homelessness in Urban America', *Urban Geography*, 15: 150–67.

VOLZKE, J. (1994*a*), 'Santa Ana Camping Law Gets Judge's OK', *Orange County Register*, 20 January, B1.

—— (1994*b*), 'Santa Ana Camping Law Is Upheld', *Orange County Register*, 15 September, B1.

WAGNER, D. (1993), *Checkerboard Square: Culture and Resistance in a Homeless Community* (Boulder, Colo.: Westview Press).

WALLACE, R., FULLILOVE, M., FULLILOVE, R., GOULD, P., and WALLACE, D. (1994), 'Will AIDS Be Contained Within U.S. Minority Urban Populations?', *Social Science and Medicine*, 39: 1051–62.

WARREN, J. (1997), 'Challenging Charity', *Los Angeles Times*, 4 March, A3.

WEIKEL, D., and MARQUIS, M. (1994), 'Programs for Poor Suffer Deepest Cuts', *Los Angeles Times*, 23 December, A1, A26.

WEINREB, L., and ROSSI, P. H. (1995), 'The American Homeless Family Shelter "System"', *Social Service Review*, 86–107.

WEISBERG, B. (1993), 'One City's Approach to NIMBY: How New York City Developed a Fair Share Siting Process', *Journal of the American Planning Association*, 59: 93–7.

WEITZMAN, B. C., KNICKMAN, J. R., and SHINN, M. (1992), 'Predictors of Shelter Use Among Low-Income, Families: Psychiatric History, Substance Abuse, and Victimization', *American Journal of Public Health*, 82 11: 1547–50.

WEKERLE, G. R., PETERSON, R., and MORLEY, D. (1980) (eds.), *New Space for Women* (Boulder, Colo.: Westview Press).

WENNERHOLM, R. W. (1994), 'Enforcement Problems and the Homeless', *The Police Chief*, 61: 13.

WEST, C. (1986), 'Populism: A Black Socialist Critique', in H. C. Boyte and F. Riessman (eds.), *The New Populism: The Politics of Empowerment* (Philadelphia: Temple University Press), 207–12.

WEST, G. R., and VALDISERRI, R. O. (1994), 'Understanding and Overcoming Obstacles to Planning HIV-Prevention Programs', *AIDS and Public Policy Journal*, 9: 207–13.

WESTON, B. (1993), 'Homeless Backers Win Round in Court', *Orange County Register*, 9 October, B5.

WILL, G. F. (1995), 'One for Santa Ana', *The Washington Post*, 7 May, C7.

WILLIAMS, F. B. (1994), 'Setting Up Homeless Camp Has Its Skeptics', *Los Angeles Times*, 15 October, B3, B8.

WILSON, D. (1996), 'Metaphors, Growth Coalition Discourses and Black Poverty Neighborhoods in a U.S. City', *Antipode*, 28: 72–96.

WILTON, R. D. (1996), 'Diminished Worlds? The Geography of Everyday Life with HIV/AIDS', *Health & Place*, 2: 69–83.

WOLCH, J. R. (1996), 'Community-Based Human Service Delivery', *Housing Policy Debate*, 7: 649–71.

—— and DEAR, M. (1993), *Malign Neglect: Homelessness in an American City* (San Francisco: Jossey-Bass).

—— and GABRIEL, S. A. (1985), 'Dismantling the Community-Based Human Service System', *Journal of the American Planning Association*, 51: 53–62.

—— DEAR, M., and AKITA, A. (1988), 'Explaining Homelessness', *Journal of the American Planning Association*, 54 4: 443–53.

—— RAHIMIAN, A., and KOEGEL, P. (1993), 'Daily and Periodic Mobility Patterns of the Urban Homeless', *Professional Geographer*, 45: 159–69.

WOLPERT, J., and WOLPERT, E. (1976), 'The Relocation of Released Mental Hospital Patients into Residential Communities', *Policy Sciences*, 7: 31–51.

WOOD, D., VALDEZ, R. B., HAYASHI, T., and SHEN, A. (1990), 'Homeless and Housed Families in Los Angeles: A Study Comparing Demographic, Economic, and Family Function Characteristics', *American Journal of Public Health*, 80: 1049–52.

WRIGHT, J. D. (1989), *Address Unknown: The Homeless in America* (New York: Aldine de Gruyter).

—— and WEBER, E. (1987), *Homelessness and Health* (Washington, DC: McGraw-Hill).

WYATT, G. E. (1991), 'Examining Ethnicity Versus Race in AIDS Related Sex Research', *Social Science and Medicine*, 33: 37–45.

YOUNG, I. M. (1990), *Justice and the Politics of Difference* (Princeton: Princeton University Press).

ZUKIN, S. (1991), *Landscapes of Power: From Detroit to Disney World* (Berkeley: University of California Press).

Index